Chemical & Microbiological Analysis of Milk & Milk Products

Chemical & Microbiological Analysis of Milk & Milk Products

Ramakant Sharma, M.Sc., Ph. D.
Incharge Bacteriologist
Bhopal Sahakari Dugdha Sangh Maryadit Dairy Plant
Habibganj
Bhopal, Madhya Pradesh (India)

International Book Distributing Co.
(Publishing Division)

Published by

INTERNATIONAL BOOK DISTRIBUTING CO.
(Publishing Division)
Chaman Studio Building, 2nd Floor,
Charbagh, Lucknow 226 004 U.P. (INDIA)
Tel. : Off. : 2450004, 2450007, 2459058 Fax : 0522-2458629
E-Mail : ibdco@sancharnet.in

First Edition 2006

Pric MAXIMUM RETAIL PRICE INCLUSIVE OF ALL TAXES Rs 995.00

ISB[illegible]1-5

Composed & Designed at :
Panacea Computers
3rd Floor, Agrawal Sabha Bhawan
Subhash Mohal, Sadar Cantt. Lucknow-226 002
Phone : 0522-2483312, 9335927082
E-mail : prasgupt@rediffmail.com

Printed at:
Salasar Imaging Systems
C-7/5, Lawrence Road Industrial Area
Delhi - 110 035
Tel. : 011-27185653, 9810064311

Late Shri Krishna Prasad Sharma   Shrimati Mugia Devi Sharma

This book is dedicated to my parents, Late Shri Krishna Prasad Sharma and Shrimati Mugia Devi Sharma without whose encouragement, inspiration and moral support it would never have been written.

Ramakant Sharma

अर्जुन सिंह
ARJUN SINGH

सत्यमेव जयते

मानव संसाधन विकास मंत्री
भारत
नई दिल्ली–110 001

MINISTER OF
HUMAN RESOURCE DEVELOPMENT
INDIA
NEW DELHI-1100 01

MESSAGE

The book entitled 'Chemical and Microbiological Analysis of Milk and Milk Products' covering all procedure by using the quality of milk and milk products are measured and its quality is ensured for market supplies. This book will be very useful to professionals, entrepreneurs, scientists, graduate & post-graduate students of food technology, microbiology, dairy technology and research scholars.

I am glad that the book will be published and will be used as a practical guide for analysis of milk and milk products to help modernize production of milk and milk products. I hope this book will be useful for both professionals and libraries and also for the students of Dairy technical institutions and employees of companies who are involved in day-to-day dairy industry.

ARJUN SINGH

RAJ BHAVAN

BHOPAL-462003

Dr. Bal Ram Jakhar

Message

From time immemorial, India's traditional foodstuff with their extraordinary variety and richeness have served people's needs for nutrition and sound health. The progress in dairying is transforming at an incredible pace form an age-old backyard vocation to a dynamic agribusiness.

Dairy products such as butter, cheese, milk powder, baby and malted food that were imported earlier are now indigenously made. The transforming factor is due to grassroots rural organization with professional management; a total commitment powered by a missionary zeal. The facets of Operation Food Programme have made all this possible.

In order to fulfil the promise of dairying, the accumulated knowledge and experience needs to be disseminated. I feel that the compilation in its present form will facilitate understanding of milk & milk products's production, processing and quality concepts and using them in addressing operational problems.

Dr. Ramakant Sharma has made a significant and notable contribution by putting large number of facts and figures on the varying aspects of milk production and processing. This blend of technical and informative detail will help better utilization of existing facilities. They are indispensable to dairymen for apprising themselves in upgrading their skills. The book **"Chemical & Microbiological Analysis of Milk & Milk products'** is valuable and will be helpful to dairymen.

I extend my warm greetings and wish all success.

Bhopal

March 07,2005

(Bal Ram Jakhar)

BABU LAL GAUR
(Chief Minister)

Governement of Madiya Pradesh
BHOPAL - 462004
March, 24, 2005

Message

I am happy to note that books on milk products, processing and quality testing have been authored.

Madhya Pradesh is fast emerging as a milk producing state. The Godan Yojana would give a fillip to the process. In this context, it is pertinent to enlighten people specially the professionals about quality norms in dairy sector.

I hope the book "Chemical and Microbiological Analysis of Milk and Milk Products" would be well received.

With best wishes.

(Babulal Gaur)

सी—11, स्वामी दयानंद नगर,
(74 बंगले), भोपाल
दूरभाषः मंत्रालयः 2441361
निवास : 2551687, 2441780
विधानसभा :244270
रीवा:07662—242355

राजेन्द्र शुक्ल
राज्यमंत्री(स्वतंत्र प्रभार)
आवास एवं पर्यावरण विभाग
मध्यप्रदेश

Message

Nowadays Indian milk and milk products are the fastest growing products in the dairy industry. Milk products offer opportunity for absorbing the growing milk surplus, generated by the Operation Flood. The Indian dairy products have not only served as a cultural link with the modern dairy industry but have also provided a technological base for diversification, export promotion and as a value added product to make the modern dairy industry economically strong to enable the milk producer to benefit from it. The book entitled "Chemical and Microbiological Analysis of Milk and Milk Products "will be very useful to achieve the desired quality of milk and milk products.

I appreciate the efforts of the book to combine the learning experiences of the dairy industry in production of milk and milk products.

I wish all success to fulfill the purpose.

(Rajendra Shukla)

रमाकांत तिवारी
विधायक

70, त्योंथर
पूर्व मंत्री, म.प्र. शासन
एवं
अध्यक्ष
म.प्र. हस्तशिल्प एवं हाथकरघा निगम
बी–25(74 बंगला), स्वामी दयानन्द नगर,
भोपाल, पिनकोड–462003

संदेश

भारत एक कृषि प्रधान देश है। यहां पर प्रचुर मात्रा में दुग्ध उत्पादन होता है। लेकिन दुग्ध तथा दुग्ध उत्पाद की गुणवत्ता विश्व व्यापार में ले जाने के लिये कोडेक्स एलिमेंनटेरियस के स्टैण्डर्ड प्राप्त करने के लिये यह पुस्तक केमिकल एण्ड माइक्रोबायोजोलिकल एनालिसिस ऑफ मिल्क एण्ड मिल्क प्रोडक्टस अति उपयोगी सिद्ध होगी।

इस पुस्तक को पढ़कर दुग्ध तथा दुग्ध उत्पाद के संसाधन करने में सहायक सिद्ध होगी। यह पुस्तक डेरी तकनीकी, फूड तकनीकी, माइक्रोबायोलोजी के विद्यार्थियों के लिये तथा शोधार्थियों के लिये अति उपयोगी होगी। डॉ. रमाकांत शर्मा के इस लोकोपयोगी प्रयास के लिये मैं हार्दिक शुभकामना देता हूँ।

(रमाकांत तिवारी)
अध्यक्ष

निवास : बी–32 बंगले
(स्वामी दयानंद नगर), भोपाल
मंत्रालय : 2441438
दूरभाष निवास : 2441224
2761659

सुनील नायक
राज्य मंत्री (स्वतंत्र प्रभार)
पशुपालन
मध्यप्रदेश शासन

संदेश

हर्ष का विषय है कि ''केमिकल एण्ड माइक्रोबायोलॉजिकल एनालिसिस ऑफ मिल्क एण्ड मिल्क प्रोडक्टस'' पुस्तक का प्रकाशन किया जा रहा है।

यह पुस्तक दुग्ध तथा दुग्ध उत्पादों को पी.एफ.ए. एवं आई.एस.आई. के मानक मापदण्डों के अनुसार दुग्ध एवं दुग्ध उत्पादों की गुणवत्ता को सुनिश्चित करने में सहायक होगी वहीं दूसरी ओर पुस्तक फूड टेक्नालॉजी, डेयरी टेक्नोलॉजी तथा शोधकर्ता विद्यार्थियों के लिये भी उपयोगी होगी।

मैं पुस्तक के लेखक डॉ. रमाकांत शर्मा के इस प्रयास की सराहना करता हूँ।

पुस्तक के सफल प्रकाशन हेतु मेरी ओर से हार्दिक शुभकामनाएँ।

Sunil

(सुनील नायक)

C– 19, Shivaji, Nagar,
Bhapal 462016
Churhat Distt. Sidhi (M.P.)
Tel. : (0755) 2555355
Fax : (0755) 2550934

AJAY SINGH 'RAHUL'
M.L.A.

Message

The book entitled "Chemical and Microbiological Analysis of Milk and Milk Products" covers the procedures of milk and milk products analysis. It is highly useful for dairy professionals, entrepreneurs, graduate and post-graduate students of microbiology, food technology, dairy technology and research scholars.

I am sure, this book will be also useful for libraries, dairy technical insitutions and employers of companies who are involved in day-to-day dairy industry.

(Ajay Singh)

रमेश चन्द राजपूत
अध्यक्ष
Ramesh Chandra Rajput
CHAIRMAN

भोपाल सहकारी दुग्ध संघ मर्यादित
Bhopal Sahakari Dugdha
Sangh Maryadit
(पी.बी.एक्स.) 587251–55
फैक्स : 0755–580896
TEL : (P.B.X.) 507251-55.
Fax " 0755-580896
भोपाल डेरी प्लांट, हबीबगंज,
भोपाल–462024
Bhopal Dairy Plant, Habibganj,
Bhopal - 462024
निवास : अमृत निवास, शुजालपुर (म.प्र.)
Resi : Amrit Niwes, Shujalpur (M.P.)
07360-42284, 42106

MESSAGE

The book entitled "Chemical and Microbiological Analysis of Milk and Milk Products" will be very useful to achieve the desired quality of milk and milk products. The book will be highly useful for modern technology, research, development and creation of infrastructural facilities and also helpful for tremendous boost to dairy development especially in the cooperative sector.

I commend the efforts of the book's author, Dr. Ramakant Sharma, to combine the learning experiences of the dairy industry in manufacture of milk and milk products with new process, technologies and modern management. I expect this effort to contribute in bringing about a new approach to raise the quality of milk and milk products.

(Ramesh Chandra Rajput)

K.K. Singh

Commissioner, Veterinary &
State Registering Authority
(M.M.P.O.)

M.P. Bhopal.

Message

I am extremely happy to know that the author, Dr. Ramakant Sharma has brought in new concepts for analysis of milk and milk products.

The book entitled "Chemical and Microbiological Anaylysis of Milk and Milk Products" offers opportunity for absorbing the growing surplus milk generated by the Operation Flood. The book will be highly useful to achieve the desired quality of milk and milk products. It will be useful for technoologists and professionals involved in research and development and creation of infrastructural facilities.

(K.K.SINGH)

Preface

The Indian Dairy Industry has made rapid progress since Independence. A large number of modern milk plants and product factories have since been established. These organized dairies have been successfully engaged in the routine commercial production of pasteurized milk and various western and Indian dairy products. Most of the supervisory and technical personnel in these dairies have had their dairy education in this country, although a few have been trained abroad as well. The author is interested in sharing his knowledge of chemical and microbiological analysis of milk & milk products with many such persons i.e. consumers, producers, graduate and post-graduate students of food technology, microbiology, dairy technology, doctors, and research scholars in the interest of our business and for the safety of our consumers. It is essential to make awareness regarding food **safety standards.**

The term quality covers physical, chemical, microbiological and safety aspects of a product. A poor quality food product will have poor market value, short shelf life and could also be a health hazard to consumers.

In most of the developed countries, the media, the consumers and the regulatory bodies have forced the producers and the suppliers to comply with the prescribed food safety standards. Whereas in India, while there are reasonably good rules, observations of these rules and the implementation of quality control measures are usually inadequate, due to many reasons. Public awareness regarding quality is poor and regulations on quality are rarely followed.

In the interest of our business and for the safety of our consumers, it is essential to maintain the appropriate quality of milk and milk products. As a member of the World Trade Organization, it has become obligatory for India to apply sanitary and phytosanitary measures while producing, processing and marketing milk and milk products and abide by the guidelines prescribed by the Codex Alimentarious Commission. This is necessary if we want to participate in the international trade of milk and milk products for ensuring better returns to our farmer members. The United Nations recommends governments all over the world to adopt the standards set up by the

Codex Alimentarious as the reference point for consumer protection, with regard to food. By harmonizing the food laws and by adopting internationally agreed standards, global commerce improves and the trade barriers diminish gradually. Through harmonization free movement of goods amongst the countries is achieved to the benefit of the farmers and the subsequent reduction of hunger and poverty.

Several international firms have started importing milk to India and have even set up their plants here.

We cannot prevent them from doing so, because of the WTO agreement. In order to safeguard the existing market and expand the business of the co-operative dairy sector, it is essential to produce superior quality products that are safe for consumers.

In order to fulfil the information appropriate to producers, consumers and milk processers, the book has been prepared. I shall welcome the comments and criticism from readers to further improve the book. There are 41 chapters and by reading this book, I hope the reader, besides basic concepts, will also gain an overview of the current status of many diverse and interesting aspects of chemical and microbiological analysis of milk and milk products.

I am thankful to Shri Shailendra Singh (I.A.S.), Dr. Rajesh Rajora (I.A.S.) and Managing Director of M.P.C.D.F., Bhopal, for their valuable expert suggestions in the respective areas.

I express a deep sense of gratitude to Dr. D.K. Singh, Consultant, (R and D) NDDB Anand and Shri Suresh D. Jaisinghani Manager (QPM) State Office, NDDB Bhopal, for their Support and encouragement.

I am highly indebted to Shri Vinod Kumar Shukla, Managing Director, Constructing Committee Company, Bhopal, Shri Triveni Prasad Mishra, Advocate, Supreme Court New Delhi, Shri Rakesh Kumar Shukla, Advocate, Bhopal and Shri Dwarika Prasad Shukla, Administrative Academy, Bhopal, for their help and encouragement in this endeavor.

I would like to express my deep gratitude to Dr. A.K. Nigam General Manager, Bhopal Sahakari Dugdha Sangh, Bhopal, for offering invaluable suggestions regarding the book. I am thankful to Shri M.E. Khan, Manager, Field Operation, B.S.D.S., Bhopal, for continued help and encouragement. I am highly indebted to Dr. D.S. Bisen, O.S.D.

Administration, for offering invaluable suggestions. I also express my deep gratitude to Shri D.N. Singh Plant operation Manager, Shri R.P. Bilongu, Manager Marketing, and Shri M.C. Bansal, Deputy Manager, Finance, B.S.D.S, Bhopal, for their support in this endeavor.

I am highly indebted to Shri D.K. Roy Joint Director, Shri C.D. Khan, Walkar Manager Quality Control, Shri S.K. Ranawat, Joint Director, and Shri S.C. Mandake, Joint Director, M.P. C.D.F. Bhopal for their encouragement and support.

I am thankful to Dr. R.R. Singh G.M. G.S.D.S. for his support and encouragement in this endeavor.

I am also thankful to Shri N.S. Pawar G.M. U.S.D.S and Dr. S.N. Patel, Incharge G.M. U.S.D.S, Shri Alam Khan G.M. I.S.D.S. and Shri G.R. Daruwala M. (PO) I.S.D.S. for their support and encouragement.

I express a deep sence of gratitute to Shri Sudhir Kapoor (retired Director) and Shri R.S. Dwivedi (Retired Joint Director) M.P.C.D.F. Ltd. for their continued support and encouragement.

I am thankful to Shri S.B. Joshi, Project Officer, Rewa and Shri G.K. Bhatnagar, Project Officer, Jhabua, for their support and encouragement.

I extend my grateful thanks to Dr. (Mrs.) Kiran Singh, HOD Deptt. of Microbiology, Barkatullah University, Bhopal, Dr. Krishna Jha Principal Scientist (Microbiology), CIAE, Nabibagh, Bhopal, Dr. R.K. Tenguria, Asstt. professor and head of Deptt. of Microbiology, M.V.M. College Bhopal, Dr. Ashwani Wanganeo, (Reader), Department of Limnology, Barkatullah University, Bhopal, Dr. S.S. Khan, professor and head of Deptt. of Microbiology, Safia College, Bhopal, Dr. Vivek Mishra, Assist. Professor, Deptt. of Microbiology, Saifia College, Bhopal, Dr. V.P. Dwivedi, Barkatullah University, Bhopal, Shri S.K. Choubey, Biologist, PHE Deptt. Satpura Bhawan, Bhopal, Dr. Chandrika Prasad Dwivedi, Lecturer, Rewa and Dr. Santosh Kumar Mishra, Sr. Ayurved health officer, Rewa for their help and encouragement.

I thank the following officers for their help and encouragement: Shri R.K. Singh, AM (MIS)/ MR, Shri M.R. Mahato, AM(QC), Shri S.K. Bhardwaj AM(QC), Shri Awadhesh Khare (AMP), Shri Basant Chaudhari AM(E), Shri Anil Kashiv AM (IM), Shri Ashok Khare Supdt. (QC), Shri Ravi Saxena incharge OSD, Shri Syam Gupta

Supdt.(QC), Smt. Snehalata Sinha Supdt.(QC), Shri M. Purohit Processing Supdt., Shri B.P.S. Rathore Marketing Suptd., Smt. Sudha Shrivastava Sr. Assist. (QC), Smt. Alka Bajpai Sr. Assist. (Q.C.), Shri B.P., Jain Sr. Assist. (Q.C.), Shri S.K. Shrivastava Sr. Asist. (Q.C.), Shri Rajeev Das Jr. Assist. (Q.C.), Shri Syam Shukla Sr. Assist. (Store), Shri Raj Kumar Malviya Technician Electrical, Shri Bharat Singh Chauhan V.E.O., and all the processing supdt., all the quality control staff, all the engineering staff, all the marketing staff, all the administrative staff, all the field staff, and all the members of Dairy colony for their support and encouragement.

I express a deep sence of gratitude to Late Shri Ramgopal Payasi (my grandfather), Shri Ramji Sharma (my elder uncle), retired assistant engineer. Shri V.D. Sharma (my elder uncle), retired additional executive engineer, Dr. R.D. Sharma, senior veterinary officer (my younger uncle), Dr. S.K. Payasi, senior ayurved health officer (my elder brother) and Shri Umakant Mishra (my younger brother), Shri Devendra Kumar Sharma (my younger brother) late Shri Vijay Kumar Payasi (my younger brother) and all Payasi family for their help and encouragement.

I extend my grateful thanks to Shri Kamata Prasad Shukla (assistant grade-IInd) M.P. Vidhan Sabha, Shri Pushpendra Shukla (assistant engineer) G.E.I. Hamon Industries Ltd. Govindpura, Bhopal, Shri T.P. Mishra (senior accountant), Forest Department, Shri L.P. Mishra (staff officer) B.H.E.L. Bhopal, Shri Ramji Mishra (office supdt.) Barkatullah University Bhopal, Shri Ramraj Tiwari (foreman) H.E.E.G. Mandideep, Shri L.P. Kurmbansi M.P. Board Office Bhopal, Shri Rajeshwari Prasad Mishra (staff officer) M.P. Vidhan Sabha, Shri J.R. Pande (office suptd.) Pathya Pustak Nigam Bhopal for their help and support.

I express a deep sense of gratitude to Shri Harihar Prasad Tiwari (my sisters' father-in-law), Shri Murlidhar Pande (my elder father-in-law), retired principal, Shri Suryadhar Pande (my father-in-law), Shri Surya Mani Pande (my elder father–in-law), Shri Tulsi Prasad Tiwari (my brother-in-law), Smt. Prabhavati Tiwari (my sister) & Shri Mukund Lal Pande (my brother-in-law) for their encouragement & support in this endeavor.

I extend my grateful thanks to Shri Tezpratap Shukla, retired head master, and Shri Siddharth Shukla, marketing officer, D.S.F.S. Advisory Services Pvt. Ltd. Ahmedabad.

I am highly indebted to Shri C. Anandan and Mrs. Renuka Anandan for their help and encouragement. I am indeed grateful to Shri Ashok Kumar Kumbhare, Kothari Photostat & Typing Centre, Bhopal, for typing the manuscript diligently and accurately. Thanks are due to all my friends and well wishers for their continued support in this endeavor.

I am grateful to my wife, Mrs. Savitri Sharma, for understanding and encouragement. I am thankful to my daughters, Ku Shailja Sharma and Ku Swapna Sharma for their help in preparation of the manuscript and proofs reading. I would also like to thank the publishers who undertook the task of publishing the book.

February- 2005

Dr. Ramakant Sharma
Incharge Bacteriologist
Bhopal Sahakari Dugdha Sangh
Habibganj, Bhopal

Contents

Chapter 1

Sampling of Milk

1. **Introduction** – Sampling of milk shall be done by an experienced person. It is not possible to lay down a single sampling procedure for milk which will be applicable in all cases. The method of sampling will vary according to the purpose for which the sample is collected and the tests which are to be carried out. Proper sampling, however, requires the most careful attention to the recommended procedures in this standard.

1.1 Samples may be required for chemical or bacteriological examination. All precautions shall be taken to prevent contamination and adulteration.

1.2 For chemical examination, the sampling equipment shall be clean and dry.

1.3 Samples for bacteriological examination shall be collected by a person trained in the technique of sampling for bacteriological work.

1.3.1 For bacteriological purposes, all equipment including plungers, sample bottles and rubber stoppers shall be sterile and the samples collected under aseptic conditions. Equipment shall be sterilized by one of the following methods:

(a) Heating in a hot air oven for not less than 2 hours at 160ºc, or

(b) Autoclaving for not less than 15 minutes at 120ºc.

NOTE-I Under field conditions, equipment may be sterilized by immersion for at least 5 minutes in boiling water. Equipment treated by this method shall be used immediately.

NOTE-II Rubber stoppers shall be sterilized in an autoclave as in (b). Treatment by immersion in boiling water for not less than 10 minutes would be satisfactory if they are used immediately.

1.1.4 The sample collected for chemical analysis should be representative of the entire batch of milk that is being sampled. Since milk fat is of lower density than the other constituents of milk, it tends to rise to the surface. Thorough mixing of milk with a proper instrument which will reach the entire depth of the liquid is essential to ensure a representative sample of the entire batch. In small batches, it should be possible to accomplish mixing by pouring the entire quantity of milk from one container to another, three or four times. Larger batches of milk shall be thoroughly agitated by a hand stirrer or by mechanical means. Milk churns easily at 26.5ºc to 29.5ºc and agitation near this temperature shall be avoided.

2. **Sampling from Individual Container** - Pour the milk from one container to another, three or four times. Where this is not practicable, mix thoroughly with a plunger. In mixing the milk, the plunger shall be allowed to fall to the bottom of the container and brought to the top of the milk as rapidly as possible not less than 10 times. The position of the plunger shall also be moved from place to place to ensure that the whole of the milk at the bottom of the vessel is thoroughly agitated and mixed with the upper layer. Any milk fat adhering to the neck and under the shoulder of the can shall be well mixed with the remainder of the milk. After thorough mixing, a sample shall be drawn immediately.

3. **Sampling from Several Containers** - The samples shall be taken after pouring the contents of the containers into a vat and mixing. When this is not possible, a composite sample is taken in the following manner from the containers after milk has been agitated and mixed. First, the milk shall be distributed as equally as possible among a number of containers. The cans shall not be filled, but the same quantity shall be placed in each. After mixing the contents of each can thoroughly, an equal volume of milk shall be taken from each. These portions shall

be placed in another vessel, thoroughly mixed as described in 2 and a sample then taken.

3.1 Alternatively, where facilities exist for accurate measurements, a composite sample may be obtained by taking the same proportion of the milk therein from each container in a consignment after thorough mixing, collecting this in another vessel and taking a sample as described in 2.

4. **Sampling Bulk Units** – When milk of uniform quality is supplied in bulk units (for example, cans filled from storage tanks), the number of random units to be sampled shall be as follows:

Total number of units	Number of units to be selected
1	1
2-5	2
6-20	3
21-60	4
61-100	5
Over 100	5 plus one for each additional 100 units or fraction thereof.

4.1 The testing laboratory, may, within its descretion, instruct the person who draws the sample to submit;

(a) Separate samples from each unit selected, or

(b) One or more composite samples consisting of aliquot portions from each unit selected.

The latter course should only be applied where the product is likely to be of fairly uniform composition, for example, where the consignment to be sampled is produced from a quantity of properly bulked milks and where variations in composition from unit to unit are, therefore small. Where there is a possibility of wide variations between different units, for example, a consignment of milk from an individual producer, every selected unit shall be separately sampled.

5. **Sampling from storage tanks and rail and road milk tankers** - The method of sampling of milk from storage tanks and rail and road tankers is largely governed by storage/transport condition. It is, therefore, difficult to lay down any rigid procedure for the sampling, but the following is recommended:

   (a) In all cases, the milk in the tank/tanker shall be thoroughly mixed by a sufficiently large plunger, a mechanical agitator or by compressed air, the uniformity of the samples being determined, when necessary, by mixing till such time as complete agreement is obtained between samples taken at the manhole and at the outlet cock in respect of fat and total milk solids.

   Note- When a plunger is used for mixing the milk in rail or road milk tankers, a convenient and satisfactory method is to insert the plunger in the manhole, the operator sitting astride or standing on top of the tanker. The plunger is thrust forward and pulled back, thrust backwards and pulled back. The cycle of operations should be repeated for at least 15 minutes.

   (b) After proper mixing of the milk, the sample may be taken from the tank, removed through the stop-cock in the tank door, or from a valve on the discharge line from the tank as it is being emptied.

6. **Composite milk samples for fat test** - Suppliers of milk are often paid for milk on the basis of fat test. The determination of fat contents of the suppliers, daily deliveries is laborious and expensive. Composite samples of the suppliers milk are taken over a period and then tested. The volume of the individual composite sample shall be not less than 175 ml and it shall be collected during the agreed period by placing into the patron's composite sample bottle proportionate amounts of the suppliers' daily delivery. For preserving the composite sample, 0.1 ml of 36 per cent formaldehyde for 25ml of milk may be used. The bottle containing the composite milk sample shall be tightly, stoppered to prevent evaporation and kept in a locker, away from light, till required for analysis. The sample shall be analysed on the same day as the last portion of milk is

transferred to the composite sample bottle.

Note: Each time when fresh sample of milk is added, the sample shall be mixed by rotating the bottle to prevent the formation of solid cream layer or cream plug.

7. **Appliances for sampling** – The following appliances are required for sampling:

7.1 Plungers,

7.2 Sampling dippers and

7.3 Sampling tube.

They shall preferably be made of stainless steel, but adequately tinned iron may also be used. If solder is employed, it shall be capable of withstanding a sterilizing temperature of 180ºc. All surfaces shall be smooth and free from crevices or projections.

7.1 Plungers –Plungers shall have sufficient area to produce adequate disturbance of the product and sufficiently light in weight for the operator to be able to move them rapidly through the liquid. In view of the differing shapes and sizes of containers, no specific design of plunger can be recommended for all purposes. A form of plunger recommended as being suitable for the mixing of milk in buckets or in cans consists of a disc 150mm in diameter, perforated with six holes each 12.5mm in diameter on a pitch circle of 100mm diameter, the disc being fixed centrally to a metal rod, the other end of which forms a loop handle. The length of the rod, including the handle, should be approximately one metre. A suitable plunger for use with road and rail tanks has a rod not less than 1.8m in length and is fitted with a disc 300mm in diameter perforated with twelve holes each 30mm in diameter on a pitch circle of 225mm diameter.

7.2 Sampling Dippers – Sampling dippers shall be fitted with a solid handle at least 150mm long. The capacity of the sampling dipper shall be not less than 80ml. It is advantageous to have a dip and have the handle bent over. The tappered form of the cup permits nesting of the sampling dippers. The arc of the inside bottom corner of the sampling dipper shall be defined to assure proper cleaning. The body of the sampling dipper shall be of one-piece construction with no seams, overlaps, revets or sharp corners.

Note:- When approximate quantity of milk is to be measured, dipping measures of convenient quantities may also be used.

7.3 **Sampling Tube-** Straight seamless metal tube about 600mm long, 6mm inside diameter and about 1.6mm thick may be used for sampling where convenient.

8 **Sample Bottles** - The sample bottles shall be made of good quality glass suitable for sterilization. The sample bottle shall be wide mouthed, round with sloping sides on the pattern of the milk bottles. Bottles used for collecting samples for chemical analysis shall be provided with well fitting caps or bark corks. Bottles for collecting samples for bacteriological examination shall be glass stoppered

8.1 The capacity of the sample bottle shall be 100, 150 or 250ml. The size of the sample bottle selected for taking a sample shall be such that after containing the quantity of milk required for analysis, only a small space would be left for efficient mixing of the sample as a larger space would allow the fat to churn during transit.

8.1.1 When samples are collected for bacteriological examination, it is desirable to avoid air space by filling the bottles to the top leaving, however, sufficient space to allow for expansion of the rubber stopper.

8.2 Bark corks shall not be used for closing milk sample bottles for bacteriological examination.

8.3 Sample bottles containing milk which are to be examined for flavour subsequently shall be closed with grease-proof, non-absorbent stopper so that no deleterious odour or taste is imparted.

Note: Alternatively, for collecting samples for chemical analysis suitable plastic bottles of above capacities may also be used.

9 **Labelling of Samples** – Each sample container shall be sealed air tight after filling and marked with particulars regarding the purpose of sampling, the name of the supplier or other particulars of the stock, the date and time of sampling, the nature of preservatives, if any, added and any other relevant information. Samples for bacteriological examination shall be marked distinctively.

10 **Transport of Samples**– Milk samples which are to be examined for flavour shall be protected from light and shall not be exposed to odours which may be absorbed during transport.

(a) Samples for Chemical Examination –

It is desirable that samples of milk for chemical examination are delivered for testing on the same day they are taken. The samples shall be stored in a refrigerator at a temperature of 0 to 5ºc. Where this is not possible, adequate precaution shall be taken to prevent deterioration and exposure to high temperature and light during transit. In some cases formaldehyde may be added as a preservative to prevent deterioration, provided it does not interfere with the subsequent analysis. When formaldehyde has been added, this fact and the quantity added shall be indicated on the lable.

Note: If any sample is to be used for cryoscopic examination, mercuric chloride shall be permissible as a preservative provided the bottle is properly labelled as 'POISON'.

(b) If the tests are to determine the bacteriological quality, the milk samples shall be chilled immediately and maintained at a temperature not exceeding 4.5ºc. If the interval between sampling and examination exceeds 4 hours, the time of sampling and examination should be recorded on the analytical report. Generally the samples should be examined within 4 hours of collection. The results of analysis of any sample, the temperature of which has exceeded 7.0ºc during a storage period of 4 hours, may be unreliable. At a storage temperature of 0.0 to 4.5ºc no detectable increase in bacterial counts will occur within 24 hours.

Chapter 2

Organoleptic test and Temperature

1. **General** - Judging the quality of milk by its taste and smell requires considerable skill which could only be acquired by practice. Organoleptic tests are used in all dairies and an experienced person can pick out bad samples with a high degree of accuracy.

2. **Adopt the following procedure on the receiving plateform -**

   2.1 Smell the milk in the container immediately after removing the lid. In case of foul or abnormal smell, reject the milk or hold over for subjection to confirmatory test.

   2.2 Observe the colour of the milk. If abnormal in colour, it should be regarded with suspicion.

   2.3 Examine the milk for the following taints -

   2.3.1 Those due to developed acidity. This is the most important factor to be examined when grading milk by organoleptic test.

   2.3.2 Those due to feed, or exposure of milk to the atmosphere of the stable.

   2.3.3 Extraneous matter which might gain access to milk after milking.

   2.3.4 Oxidized flavour due to exposure of milk to light or metallic contamination from untinned containers.

3. **Temperature** - Determine the temperature of milk with a standard thermometer. Bulk raw milk, when received from a chilling station in the factory shall not have a temperature more than 7ºC.

Chapter 3

Determination of pH

1. **General** - The pH value or hydrogen ion concentration gives a measure of the true acidity of milk. The relationship between pH and acidity of milk is only approximate. In normal cow and buffalo milk, the pH ranges from 6.6 to 6.8. The value is reduced by the development of acidity. On the other hand, the pH value of milk from animals suffering from mastitis is alkaline in reaction, the value being over 7.0. The pH test is mainly used for the detection of abnormal mastitis milk.

2. **Indicator Strips** -The pH of milk may be determined rapidly by using the indicator strips. Indicator paper strips or discs are made by soaking strips of absorbent paper in a suitable indicator and drying them. A rough estimate of pH is obtained by dipping a strip of the prepared paper in milk and observing the colour. Bromocresol purple (pH range 5.2 to 6.8 - colour changes from yellow to purple) and bromothymol blue (pH range 6.0 to 7.6 -- colour changes from straw-yellow to bluish-green) are commonly used as indicators. Both narrow and wide range readymade indicator papers are available over the pH range 2.0 to 10.5.

   NOTE:-Indicator paper strips shall always be kept in a closed glass bottle and in a dry condition.

3. **Interpretation** - In normal milk the pH is well below 6.9. On an average cow milk gives a pH of 6.6 and buffalo milk 6.8. Milk of pH over 6.9 should be regarded with suspicion as indication of some diseases of the udder or of late lactation milk.

Chapter 4

Clot on Boiling (COB)

1. **Introduction** –Clot on Boiling (COB) is used as rapid platform test for accepting milk at collection and chilling centres and the dairy plants. This is a quick test for determining the quality of milk and to find out its suitability for pasteurization, boiling etc. This test also indicates the presence of developed acidity in milk. Milk showing COB positive (+ve) test should be rejected or handled separately. The COB (+ve) milk gets curdled during heat processing.

2. **Precautions** –

   2.1 Use clean and dry test tubes for the test.

   2.2 Do not heat milk on a direct flame.

3. **Materials Required** –

   3.1 Test tube - 15.0 x 1.9cm, preferably with a mark of 5 ml.

   3.2 Pipette - 5ml.

   3.3 Boiling water bath.

   3.4 Milk

4. **Procedure** –

   4.1 Examine milk for flakes, if any.

   4.2 Take approximately 5ml milk in a test tube.

   4.3 Dip the test tube in a boiling water bath.

   4.4 Hold the test-tube in the bath for 5 to 6 minutes.

   4.5 Remove the test tube from the bath and tilt it to an almost horizontal position.

   4.6 Examine the side of the test-tube and the film of milk for the presence of clots or precipitated milk particles.

5. **Observations** – Examine the side of the test tube and the film of milk for the presence of any precipitated milk particles. The appearance of clots indicates positive (+ve) COB test.

6. **Interpretation** – The '+ve' COB test generally indicates that milk has developed acidity above 0.17 percent in the case of cow's milk and 0.2 percent in case of buffalo's milk. Such milk is unfit for processing and distribution as liquid milk.

Chapter 5

Alcohol Test

1. **Introduction** - The Alcohol test is used for rapid assessment of stability of milk to processing, particularly for condensing and sterilization. The alcohol test is useful as an indication of the mineral balance of milk and not so much as an index of developed acidity. The test aids in detecting abnormal milk, such as colostrum, milk from animals in late lactation, milk from animals suffering from mastitis and milk in which the mineral balance has been disturbed.

2. **Materials**

   2.1 Apparatus -

   2.1.1 Test-tubes - 150 x 19mm, preferably with graduation marks at 5 and 10ml.

   2.1.2 Measure for Alcohol - for 5ml.

   2.2 Reagent -

   2.2.1 Ethyl Alcohol - 68 per cent by weight or 75 per cent by volume (density 0.8675 g/ml. at 27ºc).

3. **Procedure** -Place 5ml. of milk in a test tube and add an equal quantity of alcohol. Mix the contents of the test tube by inverting several times. Note any flakes or clot. The presence of a flake or clot denotes a positive test.

4. **Interpretation** - A negative test indicates low acidity and good heat stability of the milk sample. Milk showing positive test is not considered suitable for the manufacture of evaporated milk which has to be sterilized to ensure its keeping quality.

Chapter 6

Sediment Test

1. **General** – Sediment test on raw milk reveals the extent to which visible insoluble matter has gained entrance to the milk and the extent to which such material has not been removed from milk by single service strainers. Despite the limitations as to the interpretation that may be applied when visible sediment is or is not detected in milk, the sediment test presents a simple, rapid and a quantitative measure of indicating the cleanliness of milk with respect to visible dirt. The test is carried out by allowing a measured quantity of milk to pass through a fixed area of a filter disc and comparing the sediment left with the prepared standards.

2. **Apparatus** –

   a) Sediment Tester – with filtering surface 25mm in diameter.

   b) White Lintine Cotton Discs – 32mm in diameter, exposed filtration area 28mm in diameter.

   c) Sampling Dipper – of 500ml. capacity for sampling from milk cans or weigh vats.

   d) Sieves – two, one coarse corresponding to 850-micron IS Sieve and the other fine corresponding to 425 micron IS Sieve [See IS : 460 – 1962 'Specification for Test Sieves (revised)'].

   e) Sediment Disc Ratings – showing 0.0, 0.2, 0.5, 1.0, 2.0, 3.0 mg. sediment or higher concentration as required per 500ml of milk.

   2.1 Since sediment encountered in different localities may vary in composition, density, colour and other physical characteristics thus giving rise to variations in the appearance of the discs, it is recommended that the standard sediment discs be prepared as follows:-

Make a uniform mixture of oven-dried (100ºc) materials which meet the following screening specifications :

- Cow or buffalo dung passing through fine sieve [See 2(d)]	53 Parts
- Cow or buffalo dung passing through the coarse sieve but retained on the fine one.	10 parts
- Garden soil passing through the fine sieve	27 parts
- Charcoal passing through the fine sieve	8 parts
- Charcoal passing through the coarse sieve but retained on the fine one	2 parts
Total -	100 Parts

Accurately weigh 0.1g of the above mixture and transfer to a 1000 ml flask using 50 per cent sugar solution to wash all fine particles down into the flask. Make the volume upto the mark with more of the sugar solution after most of fine particles have been wetted by shaking the half-filled flask thoroughly several times. After the volume is made upto the mark, shake the contents of the flask vigorously every 5 minutes for sufficient time (for thirty minutes to one hour) to saturate particles thoroughly when particles have been thoroughly wetted, it will be noted that the sugar solution will hold them evenly in suspension and the mixture is ready for use in making the standard discs.

On the basis of 0.1g per 1000 ml, 10ml. of the sugar solution contains 1mg of the sediment. Make test discs with one of the usual sediment testers using varying volumes of the sediment suspension. Place 150 ml or more of filtered skimmed milk in the sediment tester and add varying volumes of the sediment suspension. After forcing the milk through the disc, run through a small quantity of filtered skimmed milk to obtain a more even distribution of the sediment on the disc.

Remove the discs from the tester, mount them permanently on a stiff paper, allow to dry and then make permanent by spraying with a strong disinfectant, such as corrosive sublimate. Below each mounted standard disc on the paper, note the quantity of dried material that the dirt or filth on the disc represents.

3. **Procedure** – Take a milk sample from well-stirred cans or vats of milk with the help of the sampling dipper. Measure the quantity of milk used with reasonable accuracy. Filter the milk through a properly adjusted firm lintine cotton disc (rough side facing milk) held in the sediment tester so that a filtration area of 28mm in diameter is exposed. Compare the sediment disc with the prepared sediment standard discs and record the sediment score.

4. **Interpretation** – For the purpose of comparison, it is convenient to use about five prepared standard discs so as to classify the milk with respect to its sediment content in accordance with the specific requirement of the dairy or the milk collection depot. For the former, five discs showing 0.0, 0.2, 0.5, 1.0 and 2.0 mg may suffice. Under rural conditions discs showing 0.0, 0.5, 2.0, 5.0 and 7.0 mg sediment may be more convenient to start with. In either case, no attempt shall be made to estimate the degree of sediment in milk in more than five classes, for example, excellent, good, fair, bad and very bad.

No attempt shall be made to grade as sediment any hair, flies, pieces of hay or straw or any large particles of dirt. These shall be reported separately.

The presence of appreciable sediment in unprocessed milk supplies indicates careless or insanitary dairy farm practice. However, the lack of sediment is not always indicative of ideal conditions since visible sediment may be readily removed by straining at the dairy farm. Milk that has been divested of visible sediment by straining retains essentially the bacterial contamination incidental to the original sediment.

Chapter 7

Ten-Minute Resazurin Test

1. **Introduction** :- This test is intended as a rapid method of detecting milk supplies of poor keeping quality on the receiving plateform. Resazurin is an oxidation-reduction indicator which is blue in the oxidised stage and upon reduction due to bacterial activity or other causes it is first turned irreversible into a pink compound "resorufin" and then into the colourless 'dihydroresorufin'. The second change is a reversible reaction. During the first of reduction from resazurin (blue) to resorufin (pink) in milk distinct intermediate colour shades are developed which can be matched with standard colour discs in a comparator. During the second stage the pink colour fades out at a fast rate and the milk eventually turns white with a narrow pink band on the surface. The rate of reduction of resazurin is governed by the extent of bacterial activity in milk and can be measured by recording the colour changes at any time. This principle forms the basis of the ten-minute or one hour resazurin reduction tests for judging the bacteriological quality of milk. Resazurin is also susceptible to the reducing action of leucocytic cells.

2. **Procedure** :-

Draw milk sample from a can

↓

Then transfer 10ml in a sterilized test tube

↓

Then add 1ml bench resazurin solution (.0.05%)

↓

↓

Insert the stopper and mixed by inverting

↓

Placed in water bath maintained at 37.5ºC

↓

At the end of 10 minutes remove the tube from water bath

↓

Place the tube in the right section of the comparator.

↓

Keep Control milk tube in the left section of the comparator to compensate for the natural colour.

↓

The standard resazurin disc is revolved until the sample is matched and the corresponding disc reading noted. When the colour falls between two discs numbers it shall be recorded as half value.

3. **Results** :- Results and classification of the samples according to the following standards are given in table - 1

Table-1

Sr. No.	Resazurin Disc Reading	Keeping quality	Remarks
1.	6, 5 or 4	Satisfactory	Accept the milk
2.	3.5 to 1	Doubtful	Requires further examination
3.	0.5 to 0	Unsatisfactory	Reject the milk

Chapter 8

Alizarin-Alcohol Test

1. **Introduction** : - The stability of milk to alcohol or high temperature is considerably affected by (a) developed acidity or sweet curdling as results of bacterial growth. (b) disturbance in normal salt balance and (c) abnormal chemical composition (e.g. colostrum, late lactation and mastitis milks). The alcohol test is, therefore, used to assess the stability of milk to heat-processing particularly for condensing and sterilization. Addition of alizarin alongwith alcohol helps in finding out whether milk is acidic or alkaline.

2. **Materials :**

   2.1 Test tubes (15x1.9cm) preferably with graduation marks at 5 and 10ml.

   2.2 5ml pipettes.

   2.3 Alizarin-alcohol solution-0.2 percent of alizarin in ethyl alcohol 68% by weight or 75% by volume-density 0.8675 g/ml at 27ºC.

   2.4 Sample of milk.

3. **Procedure :-**

   3.1 Transfer 5ml of milk to a test tube and an equal quantity of alizarin-alcohol solution.

   3.2 Mix the contents of the tube by inverting several times.

   3.3 Examine for the presence of flakes or clots and also note whether the flakes are small or large.

   3.4 Observe the colour of the mixture.

4. **Interpretation :**

   The quality of the milk is judged on the basis of the formation of flakes and clots and also the colour of the mixture in the following manner. Presence of flakes or clots (with or without development of acidity) indicates poor heat stability and unsatisfactory quality of milk. The results are given in table-1

Table-1

S.No.	Presence of flakes or clots	Colour of mixture	Reference regarding heat-stability	Quality
1.	Negative	Lilac or pale red	Good (Low acidity)	Satisfactory
2.	Positive	Lilac pale red	Poor (sweet curdling)	Unsatisfactory
3.	Positive	Violet (Alkaline)	Poor (late lactation) or mastitis	Unsatisfactory
4	Positive	Brown (Acidic)	Poor (Developed acidity 0.1-0.2%)	Unsatisfactory
5.	Large flakes	Yellow (Highly acidic)	Poor (Developed acidity more than 0.2%)	Unsatisfactory

Chapter 9

Direct Microscopic Count

1. **Introduction** : - The direct microscopic method consists of examining, under a compound microscope, stained films of a measured volume of milks 0.01ml spread on glass slides over a specified area on a glass slide ie one sq.cm. Counting the number of bacterial cells or clumps of cells in a microscopic field and calculating the number of cells or clumps per ml of milk. It enables the rapid estimation of the total bacterial population of a sample of milk and also reveals useful information for tracing the sources of contamination in milk.

   This method is not suitable for examination of pasteurised milk (where dead cells which take stains are also counted). The ratio of the standard plate count to direct microscopic count has been reported to be 1:4.

   This method is used for the screening of raw milk supplies on the receiving platform and for bacteriological grading of milk. The microscopic appearance (types and arrangement of cells) of the milk film will give indication of any udder infection as well as the cause of high counts due to utensil contamination or inadequate cooling. Estimate of bacterial numbers by this test would be more accurate in the case of poor quality milk having high bacterial counts.

2. **Materials:-**

   (i) Clean grease-free slides.

   (ii) Capillary pipettes (Breed's) calibrated to deliver 0.01 ml of milk.

   (iii) Needle with bent point for spreading milk.

   (iv) Compound microscope.

   (v) Stage micrometer slide ruled in 1mm.

(vi) Newman's Stain.

(vii) Milk sample.

3. **Procedure :**

3.1 Determination of microscopic Factor:-

Place the stage micrometer on the stage of the microscope and focus on the scale first with the 16mm objective and then with the oil immersion objective. Move the stage micrometer until one end of the scale is at the edge of the microscopic field. Count the number of small divisions (0.01 mm each) in the maximum diameter of the field and thus determine the diameter of the field. The diameter ranges from 0.146 to 0.206 mm depending on the length of the draw tube of the microscope. Usually American microscopes are adjusted to a draw tube length of 160 mm and European microscopes to 170mm. The area of the field is given by the formula p r^2 or 3.416x r^2 where r is the radius (half of the diameter) of the microscopic field. The microscopic factor (MF) is given by the formula.

$$MF = \frac{\text{Area of smear (100 sq.mm)}}{\text{Area of Microscopic field}} \times \frac{1}{\text{Volume of milk (0.01 ml)}}$$

$$= \frac{10000}{3.1416 \text{x } r^2}$$

3.2 Preparation of milk smear:-

Mix the sample of milk thoroughly by shaking. Rinse the Breed's pipette thoroughly in sterile water (25ºC to 35ºC) between samples. Dip the tip of the pipette into the sample of milk and draw in and expel several times to remove traces of rinse water. Draw milk into the pipette above the graduation mark, wipe the exterior of the pipette with clean blotting paper or cloth and adjust the volume of the sample to exactly 0.01 ml. mark. Touch the tip of the pipette to the centre of a one square centimeter area on a slide and expel the entire volume of milk. With a flamed bent point needle spread the portion of milk uniformly over the centre of one square centimetre area on the slide. Wipe needle between samples on a clean dry tissue or towel.

After spreading, dry the smears at 40-45ºC within 5 minutes on level surface protected from dust and insects. Rapid heating may cause the film to crack and peel out during later treatments.

3.3 Staining the films:-

After the films are dried, dip the slides in Newman's stain (in a jar) for ½ to 1 minutes. The composition of the Newman's stain is intended to remove fat, fix the cells and stain the organisms in one single operation. The slides may be gently washed to remove excess stain by allowing water to run over from one end to the other end. If the smear is overstained, rinse in water and decolourise with alcohol slightly. The back-ground will be faintly blue while the cells will be stained deep blue.

3.4 Microscopic examination of stained films:-

Place one drop of immersion oil on the smear and examine under the oil immersion objective. Count the single organisms or well isolated clumps of cells on a number of microscopic fields. Any isolated single cell, pair of cells or clump of cells is treated as a 'clump'. The field for counting should be selected at random and represent all parts of the film. If the average number of clumps per field is under 0.5, 0.5 to 1, 1 to 10 and 10 to 30, the number of fields to be counted will be 50, 25, 10 and 5 respectively. If the number of clumps per field is over 30 then the count is recorded as uncountable. In such cases, the milk will have to be diluted suitably and then the microscopic count determined. Calculate the average number of clumps per field and multiply by the microscopic factor (MF) to give the Direct Microscopic

Clump count per millilitre of milk.

3.5 Record the results and observations regarding the types of organisms, arrangement of cells and presence of leucocyte cells in the milk smears.

4. Interpretation

(a) The following standards are tentatively suggested for assessing the bacteriological quality of raw milk samples on the basis of Direct Microscopic clump counts are given in table-1

Table-1

Direct Microscopic Clump Counts per ml.		Bacteriological quality of milk
1)	Less than 500000	Good
2)	5,000,01 to 4000,000	Fair
3)	4,000,000 to 20,000,000	Poor
4)	Over 20,000,000	Very Poor

(b) By observing the predominating types of organisms in the milk film during microscopic examination valuable information regarding sources of contamination is obtained and given in Table-2

Table-2

Types of Organisms	Probable causes of high counts or poor quality of milk
(a) Many cocci and rods in clumps and patches	Improperly cleaned milk utensils.
(b) Excessive numbers of rod shaped bacteria and spores.	Exposure of milk to dust and dirt
(c) Large number of Cocci in pairs and short chains	Improperly cooling of milk.
(d) Large number of leucocyte cells (over 50,00,00 per ml) together with long chains of cocci	Mastitis infection

Chapter 10

Detection of Mastitis

1. General –

Mastitis inflammation of the udder, is one of the most common diseases of dairy cattle and buffaloes and it is caused as a result of infection of one or more quarters of the udder by certain species of streptococci (principally Str. agalactiae) Staphylococci and occasionally by corynebacteria, coliforms and actinomycetes. Although in some cases the causative organisms are not pathogenic to man they decrease milk yield, cause permanent damage to the injured quarters and alteration in chemical composition of milk making it unsuitable for utilisation. It is, therefore, necessary to regularly examine all milking animals and even dry cows and heifers in a herd for evidence of mastitis infection so as to give suitable medical treatment to infected cases and to adopt necessary prophylactic measures for controlling the spread of infection in the herd.

The routine diagnosis of mastitis infection is based on (i) recognition of clinical symptoms in the udder, (ii) observations of the physical and chemical changes caused in milk (iii) presence of excessive numbers of leucocyte and (iv) bacteriological examination of milk for establishing the presence of causative organisms in the suspected quarters.

Some of the simple field and laboratory tests, commonly used for detection of mastitis infection, are described in this chapter.

2. Strip-Cup Test

2.1 Introduction – This is a simple field test used for finding out the presence of fibrin, mucous and clots of milk in fore milk which is an indication of mastitis infection. Most cases of acute mastitis and 10 per cent of chronic infections are detected by this test.

2.2 Materials - Strip-cup (a piece of finely woven black cloth or screen stretched lightly over a cup and held in place by a rubber band).

2.3 Procedure -

2.3.1 Wash the first 2 or 3 streams of fore milk to fall on the black cloth or screen of the strip-cup.

2.3.2 Examine for presence of milk clots, mucous or blood cells on the cloth or screen.

2.4 Interpretation - Presence of milk clots, mucous and fibrin is an indication of mastitis infection. A negative test cannot be taken to indicate absence of any infection.

3 Brom thymol blue test -

3.1 Introduction - This test is based on the fact that milk from infected udders is usually alkaline in reaction (pH 7.0 to 7.4) and this condition can be detected by observing the colour change shown by a suitable pH indicator added to milk. The test detects about 70 percent of infected cases.

3.2 Materials.

3.2.1 Test tubes marked at 5ml level.

3.2.2 One ml pipettes.

3.2.3 Brom thymol blue solution (0.04% aqueous solution)

3.2.4 Samples of milk freshly drawn from individual quarters.

3.3 Procedure -

3.3.1 Transfer 5ml of milk into the test tube upto the mark.

3.3.2 Add one ml of the indicator solution.

3.3.3 Observe the colour of the mixture.

3.4 Interpretation - Normal milk (pH 6.6 - 6.8) shows a yellow colour when brom thymol blue is added. A greenish blue or blue colour due to alkaline reaction of milk (pH 7.0 - 7.4) indicates mastitis infection. A negative test cannot be taken as evidence of absence of infection.

Note: Milk from cows in advanced lactation also has an alkaline reaction and therefore gives a positive test.

4 **Leucocyte Count -**

4.1 **Introduction** - Milk drawn from infected quarters generally contains excessive number of leucocyte cells which can be detected by making a microscopic examination of the milk smear.

4.2 **Materials -**

4.2.1 Compound microscope and other items as given indirect microscopic count of milk.

4.2.2 Sample of milk drawn from individual quarters.

4.3 **Procedure -**

4.3.1 Prepare smears of milk on glass slides and stain with Newman's stain as in the case of direct microscopic count.

4 3.2 Examine under the oil immersion lens of a microscope and count the number of leucocyte cells in a number of fields. Calculate the number of leucocyte cells (leucocyte count) per ml of milk according to the method described in "Standard Operating Procedures for Testing of Microbiological Quality of Milk".

4.3.3 Observe the presence of any long-chained streptococci associated with leucocyte cells.

4.4 **Interpretation** - If the leucocyte count exceeds 5,00,000 per ml, it is an indication of mastitis infection. Presence of long chained Streptococci is a further evidence of infection.

Note: Early and late lactations milks may also show high leucocyte counts.

5 Resazurin Rennet Test -

5.1 Introduction - Resazurin is not reduced to any appreciable extent in freshly drawn milk containing few bacteria. If excessive numbers of leucocyte cells are present due to mastitis infection resazurin is reduced rapidly to disc two or one. Mastitis milk with abnormal salt balance takes a longer time than normal milk for coagulation by rennet. Based on these observations a combined resazurin and rennet test is used for detecting milk from infected udders.

5.2 Materials -

5.2.1 Water bath (37ºc), test tubes, resazurin solution and other accessories required for resazurin test.

5.2.2 Commercial rennet solution (diluted to 1:10).

5.2.3 Samples of milk drawn asceptically from individual quarters.

5.3 Procedure -

5.3.1 Add 1 to 2 ml of rennet solution to 100ml of bench resazurin solution. Adjust the concentration of rennet solution such that one ml of resazurin rennet solution is able to bring about coagulation of normal milk in about 30 to 40 minutes at 37ºc.

5.3.2 Transfer 10ml of milk into the test tubes, add one ml of resazurin rennet solution, insert rubber bungs and mix the contents by inverting the tubes.

5.3.3 Place the tubes in a water bath (37ºc plus or minus 0.5ºc).

5.3.4 At the end of 30 minutes examine the tube for reduction of resazurin (change of colour) and signs of clotting.

5.3.5 Incubate the tubes for another 30 minutes and then observe for signs of coagulation.

5.4 Interpretation - Reduction of resazurin to purple or pink stage (discs 2 or 1) in 30 minutes and failure to coagulate within 60 minutes are positive indications of mastitis infection.

6 **Hotis Test -**

6.1 **Introduction** - This is a simple and useful test for detection of mastitis infection due to Str. agalactiae and Staph. aureus. If these organisms are present in sample of milk, they grow and form characteristic colonies adhering to the side or bottom of the test tube after incubation for 18 to 24 hours. Further incubation may result in complete coagulation of milk. If an indicator like brom cresol purple is added to the milk, colonies of streptococci will appear yellow due to acid production.

6.2 **Materials -**

6.2.1 Sterile test tube (6" x 5/8").

6.2.2 Graduated 10ml and 1ml pipettes.

6.2.3 Incubator (37 plus or minus 0.5ºc).

6.2.4 Brom cresol purple solution (0.5% aqueous solution).

6.2.5 Samples of milk freshly drawn aseptically from individual quarters.

6.3 **Procedure -**

6.3.1 Transfer 9.5ml of milk and 0.5ml of brom cresol purple solution into the test tubes. Mix the contents well.

6.3.2 Place the tubes in an incubator (37 plus or minus 0.5ºc)

6.3.3 At the end of 18 to 24 hours, remove the tubes and observe for change of colour from the initial grayish purple, formation of agglutinated yellow or brown colonies on the sides or at the bottom of the tubes, coagulation of milk or other changes.

7 **Blood Agar Test -**

7.1 **Introduction** - Generally pathogenic strains show lysis of red blood cells, (haemolysis) although all organisms showing haemolysis may not be pathogenic. Haemolysis can be easily demonstrated by noting the zone of clearance around the colony on blood agar. This test is particularly useful for detection and isolation of mastitis organisms in milk. The lysis of blood cells is due to the action of lytic agents produced by the organisms. For example, pyogenic streptococci and staphylococcus aureus elaborate powerful lytic enzymes, producing clear and wide zones around their colonies (Betahaemolysis).

7.2 Materials -

7.2.1 Sterile petridishes, dilution flasks, pipettes (10 ml and 1 ml)

7.2.2 Nutrient agar tubes.

7.2.3 Incubator (37ºc plus or minus 0.5ºc)

7.2.4 Samples of milk freshly drawn aseptically from individual quarters of suspected animals for mastitis.

7.2.5 Defibrinated blood.

7.3 Procedure -

7.3.1 In a sterile flask containing glass beads defibrinate about 20-30 ml of rabbit, sheep or ox blood.

7.3.2 Add to melted and then tempered sterile nutrient agar in test tubes at 45ºc to 48ºc sufficient quantity of defibrinated blood so as to give about 5 per cent concentration of blood in the moisture.

7.3.3 Mix blood and agar gently by rotating the test tubes between the palms and keep the blood agar warm.

7.3.4 Prepare suitable dilutions of milk samples (from affected animals) and transfer 1ml quantities of dilution to respective petridishes. Pour 10 to 15 ml of blood agar and mix by rotary motion the dilution and blood agar in the petridishes.

7.3.5 Allow blood agar to solidify and then incubate at 37ºc for 24 hours.

7.3.6 Observe for haemolytic colonies.

7.3.7 Prepare smears of haemolytic colonies and examine the morphology by Grams staining.

8 **Interpretation** – The mastitis organisms show alpha or a weak clear zone of haemolysis, beta or distinctly clear zone of haemolysis or they may show no haemolysis. Count only those having distinct clear zones as haemolytic colonies.

Chapter 11

Methylene Blue Reduction Test (MBRT)

1. **Introduction:-** This test is based on the principle that mythylene blue (an oxidation-reduction dye or indicator) which is blue in its oxidised state, is reduced to a colourless compound (Leucoform) as a result of the metabolic activities of bacteria in milk. When a solution of methylene blue is added the organisms present in milk consume the dissolved oxygen and lower the O-R potential to a level when methylene blue and similar indicator are reduced or decolourised. The time taken for the reduction of the dye (methylene blue reduction time) is influenced by the number and types of bacteria growing in milk. The greater the number of organisms present in milk and greater their activity the more rapidly is the dye reduced. The methylene blue reduction time thus gives an indication of bacterial numbers and activity in milk. The M.B.R. test is therefore, used for (i) Judging the hygienic quality of milk and grading raw milk supplies. (ii) For assessing the probable quality of milk and (iii) for detecting post pasteurization contamination in milk.

2. **Materials:-**

   (i) Thermostatically controlled waterbath maintained at 37ºC $\pm$ 0.5ºC

   (ii) Sterile test tubes without rim (150 x 16mm) preferably with marking at 10ml.

   (iii) Sterilized rubber bungs to fit into the above test tubes. The rubber bungs together with forceps are held in boiling water for 10 minutes prior to use.

   (iv) 10.0 ml. and 1.0ml. pipettes.

   (v) Clock, watch or an interval timer.

   (vi) Forceps, beakers and flasks.

(vii) Standard methylene blue solution

(viii) Two samples of milk in sample bottles (one fresh milk and the other poor quality sample)

3. **Methylene Blue Solution:-** A standard solution of methylene blue is prepared by dissolving one tablet of approved methylene blue thiocyanate or chloride (B.D.H. or Merck) in 200 ml. of cold sterile glass distilled water in a sterile flask by gentle heating in water bath or by allowing the mixture to stand for several hours to facilitate complete solution and then adding 600ml. of sterile glass distilled water. One ml. of this solution mixed with 10ml. of milk result in obtaining a final concentration of 1/300,000 for the dye which has been found to be satisfactory for the test. The stock solution must be stored in a sterile glass stoppered amber coloured bottle in a dark place. Fresh solution must be prepared once in two months.

4. **Procedure:-**

(i) Thoroughly mix the samples of the milk.

(ii) Transfer 10ml. of each sample of milk into a test tube.

(iii) Add 1ml of the methylene blue solution to the milk in the test tubes and replace the cotton plugs with sterile rubber bungs using sterile forceps. While transferring methylene blue solution care should be taken not to contaminate the pipette by touching the milk or otherwise a fresh pipette will have to be used for transferring methylene blue solution to another tube.

(iv) Mix the dye and the milk by inverting the tubes twice.

(v) Place the tubes in the water bath.

(vi) Observe the test tubes after every 30 minutes and if there is no sign of reduction (decolourisation) the tubes are inverted once and returned to the water bath. If the decolourisation has commenced the tubes should not be inverted or shaken.

(vii) Continue the observations until the complete reduction of the dye (complete decolourisation) occurs or the formation of a persistent blue ring (0.5mm) at the top.

(viii) Two control tubes, one containing 10ml. of milk and 1 ml. of the methylene blue solution, after heating it in boiling water for 3 minutes and another with 10 ml. of milk plus 1 ml of tap water are also kept in water bath. These are required for comparing the colour changes in experimental tubes.

(ix) Record the times taken for reduction of methylene blue and calculate the results.

5. **Interpretation:-** The following standards for methylene blue reduction times are suggested as a guide for grading of raw milk supplies are given in table- 1

Table-1

Sr. No.	MBR Time (Hours)	Quality of Milk
1.	5 and above	very good
2.	3 and 4	Good
3.	1 and 2	Fair
4.	½ and below	poor

Chapter 12

One-Hour Resazurin Reduction Test for Raw Milk

1. **Introduction:-** Resazurin is an oxidation-reduction indicator like methylene blue. It is initially blue in colour and undergoes reduction in two stages. First to a pink compound (resorufin) and then to a colourless compound (dihydroresorufin) as a result of bacterial activity. The first stage of reduction is an irreversible change and the second is reversible as in the case of methylene blue. In the course of the reduction of resazurin in milk there will be a series of colour changes (from blue to lilac, mauve, purple, pink, and finally colourless which can be compared with standard colour discs in a lovibond comparator and expressed in terms of standard resazurin disc, numbers (6 to 0). The time taken for the reduction of resazurin in milk to any particular stage (disc No.) or the colour change (disc No.) recorded at the end of any particular period of incubation of the milk sample is used as a criterion of bacterial activity in milk and its hygienic quality. Taking advantage of the two-stage reduction of the indicator several procedures have been proposed for reading the end point of resazurin test and judging the bacteriological quality of milk. In the one-hour modification of the test the milk sample containing resazurin solution of the test is incubated at 37ºC and the colour changes recorded at the end of one hour are used for grading raw milk supplies. Rapid reduction of resazurin to the pink and colourless stages (disc No.1 or lower) indicates high bacterial content and poor keeping quality. Since resazurin is also affected to some extent by the reducing activity of leucocyte cells present in mastitis and late lactation milk, this test also helps in detecting such abnormal milks.

2. **Interpretation:-** The following standards are suggested as a guide for grading of raw milk supplies are given in table-1

Table-1

One hour Resazurin Disc No.	Quality of Milk
4 or higher	Good
3½ to 1	Fair
½ to 0	Poor

Chapter 13

Standard Operating Procedures for Acidity Estimation and Alkaline Phosphatase test

- All the glassware used in the analysis should have been properly cleaned (including final rinsing by double distilled water there by ensuring freedom from detergent residues) and dried before use.
- All the water used in the analysis should be double distilled water.

1. Acidity of milk

1.1 General:- The titrable acidity of milk is estimated to ascertain its keeping quality and heat stability. The acidity of milk is not a true measure of lactic acid present, but in practice, gives a good indication of quality of the milk. The titrable acidity test measures the amount of alkali required to change pH of milk from its initial value of about 6.6 to 6.8 to pH 8.33. The Phenolphthalein indicator added to milk changes to pink at the end point (pH-8.33). Thus, the titration method measures the buffering capacity of milk and not the true acidity.

Due to the buffering capacity of milk, the end point of titration is not sharp and care has to be taken to adjust the conditions to reach the same end point. Numerous techniques are available but the one being described here is the simplest.

1.2 Equipment and glassware:

(i) Magnetic Stirrer

(ii) Stirring bars

(iii) Glass beakers- 100 ml

(iv) White porcelain tile/ white paper

(v) Pipettes- 10 ml.

(vi) Pipettes - 1 ml.

(vii) Burette - 10ml or 25ml.

1.3 Reagents:-

1.3.1 Phenolphthalein Indicator Solution-Dissolve lg of phenolphthalein in 120ml of 95 per cent ethylalcohol. Add 0.1 N Sodium hydroxide solution until one drop gives a faint pink colouration. Dilute with distilled water to 200ml.

It has been observed that upon storage, the faint pink tinge is lost. When this happens, add 0.1N NaoH solution again to bring the same pink colouration.

1.3.2 **Standard Sodium Hydroxide solution 0.1 N:-** Prepare a 0.1N solution of sodium hydroxide by dissolving 4.0g of sodium hydroxide pellets in small quantity of distilled water in a flask. Make up to the mark (1 litre) with distilled water. Tightly stopper the flask and shake to mix the contents.

Determine the exact strength of the Sodium Hydroxide (0.1N) as follows:-

1.3.2.1Prepare standard (0.1N) oxalic acid: Weigh accurately 6.304 g. oxalic acid (eg. wt. 63.04). Dissolve in a small quantity (say 100ml) of distilled water. Make up the volume with distilled water to the mark (1 litre). Shake it thoroughly and store in a clean dry reagent bottle. The strength remains unchanged for a long period (over a year), if evaporation is prevented.

1.3.2.2Check the strength of NaOH Solution:- Take a clean burette and rinse it with little (5 ml or so) prepared sodium hydroxide solution, fill it with the prepared sodium hydroxide solution and clamp the burette vertically on a stand. Adjust zero reading.

Pipette 10ml of oxalic acid solution in a 100ml beaker. Add 1 ml of phenolphthalein indicator and shake it well. Titrate it with sodium hydroxide solution in the burette with continuous shaking until permanent faint pink colour persists. Note down the burette reading. Repeat the titration till at least two concordant readings are obtained. Calculate the normality of the prepared solution as given below:-

$$N2 = \frac{N1\ V1}{V2}$$

Where: N1 = normality of standard oxalic acid 0.1 N

V1 = volume in ml of oxalic acid used for titration (10ml)

N2= normality of prepared NaoH (?)

V2 = volume in ml of prepared NaOH used in titration.

1.3.3 Rosaniline Acetate Solution (Stock Solution)- Dissolve 0.12g of rosaniline acetate in approximately 50ml of rectified spirit containing 0.5 ml of glacial acetic acid. Make up to 100ml with rectified spirit.

1.3.3.1 Rosaniline Acetate Solution (Bench Solution)- Dilute 1 ml. of the stock solution to 500ml with a mixture of rectified spirit and distilled water in equal proportions by volume.

Note:- The stock solution and the bench solution should be stored in dark brown bottles securely stoppered.

1.4 Procedure:-

- Thoroughly mix the milk by pouring several times from one container to another avoiding incorporation of air bubbles.

- Bring the temperature to 27ºC.
- Take three beakers and pipette, accurately, 10ml of milk in each.
- Add an equal volume of double distilled water.
- Add 1.0ml of phenolphthalein indicator solution to two of the beakers and put the stirring bars into the beakers.
- Put one of the beakers containing phenolphthalein on the magnetic stirrer.
- To the third beaker, add 1.0ml of bench solution of rosaniline acetate and put it on the white porcelain tile/white paper.
- Switch on the magnetic stirrer.
- Titrate the contents of the beaker (on the magnetic stirrer) against standard sodium hydroxide solution added drop by drop from the burette until by comparison the colour matches the pink tint of the solution in the beaker containing the rosaniline acetate solution. The pink colour should persist for at least 15 seconds.
- The titration shall be made in proper daylight or under illumination from a daylight lamp.
- Repeat the titration for second reading using second beaker containing phenolphthalein.

1.5 Calculation:-

Titrable acidity (as lactic acid per 100 ml of milk) = 0.9VN.

Where: V = Volume in ml of the standard sodium hydroxide required for titration of 10ml of milk sample.

N= exact normality of the standard sodium hydroxide solution.

1.6 Interpretation:- The normal range of acidity of milk varies from 0.10 to 0.17 per cent lactic acid. Any value in excess of 0.17 percent can safely be reckoned as developed lactic acid.

1.7 Result:- Record the calculated result as 'per cent lactic acid (or % L.A)'.

2. Phosphatase test for pasteurized milk:-

2.1 Principle:- Phosphatase enzyme at pH 9.5 (approx) and temperature 37 degree centigrade, splits the substrate, P-nitrophenyl phosphate to give p-nitrophenol, which is yellow coloured in alkaline solution.

Phosphatase present in milk is destroyed during pasteurization. Therefore, phosphatase test is performed to determine the efficacy of pasteurization of milk in a dairy plant.

2.2 Equipment and glassware:-

(i) Balance (with a readability of 1mg.)

(ii) Water bath (constant temperature maintained at $37.5 \pm 0.5^{\circ}C$)

(iii) Pipettes - 10 ml

(iv) Pipettes - 2ml

(v) Volumetric flask - 1 litre.

(vi) Volumetric flask - 100 ml

(vii) Measuring cylinder-100 ml

(viii) Test tubes with stoppers- 10 ml

2.3 Reagents:-

2.3.1 **Buffer Solution:-** Dissolve 1.5g sodium bicarbonate and 3.5g. of anhydrous sodium carbonate in distilled water and make up to one litre with distilled water.

2.3.2 Buffer-substrate solution:- Weigh accurately 0.15g of substrate (disodium P-nitrophenyl phosphate) into a 100ml volumetric flask and make up to volume with buffer solution. The solution should be stored in refrigeration and for not more than a week. The solution should not be exposed to direct sunlight. The solution should be practically colourless at the time of use. However, at times, even freshly prepared buffer-substrate may have a very faint yellow tinge, which is acceptable,

2.4 Procedure:-

- Take two test tubes and pipette 10ml of butter substrate solution into each of them.
- Stopper and bring the temperature to 37ºC.
- Bring test milk to 37ºC
- Add 2ml of test milk to one of the test tubes containing buffer substrate. This forms the 'test'.
- Boil about 10ml of milk and cool.
- Add 2ml of test milk, which has been boiled and cooled to the other test tube containing buffer substrate. This forms the 'boiled milk control'.
- Replace stoppers and shake.
- Invert to mix the contents and incubate at 37ºC for 30 minutes in a water bath covered with a lid.
- Remove the tubes after 30 minutes.
- Visually compare the colour of the test with the boiled milk control.

2.5 Interpretation:- The intensity of colour should be same in both the test and the boiled milk control. Any yellow colour in the test, in excess of that in the boiled milk control, indicates improper pasteurization. Record the observation as 'phosphatase positive' if the test is yellow (as compared to the boiled milk control), or 'phosphatase negative' if the test is of same colour as in the boiled milk control.

Chapter 14

Determination of Specific Gravity of Milk by Lactometer and Calculation of Solid-Not-Fat of Milk

1. General

1.1 Density is defined as "mass per unit volume" The unit of density is g/ml. In the case of milk, the density may be defined as mass per unit volume of milk.

Milk drawn from the udder of an animal contains a large volume of air bubbles and the fat undergoes a gradual solidification. Due to these factors, a gradual concentration in the volume of milk takes place with a minor and slow increase in the density. The density of milk will vary with the duration and temperature of storage. This variation may be overcome by ensuring that the fat is completely in the liquid state before the density reading is taken. This is done by warming the milk.

Specific gravity of milk is the ratio of density of milk to density of water at a definite temperature. Being a ratio, specific gravity has no unit.

1.2 Solid-Not-fat and Total Solid tests are conducted for the following:-

1.2.1 To know the SNF/TS of milk as the pricing of milk is done on the basis of fat and SNF.

1.2.2 For standardization of milk for the preparation of milk products of uniform quality.

1.3 The specific gravity of milk is determined to;

1.3.1 Calculate the SNF of milk.

1.3.2 Detect the adulteration of milk with water or

skimmed milk.

1.4 The density of milk can be determined by dipping a lactometer in milk. A lactometer is a simple instrument calibrated to give absolute lactometer readings in relation to water. From these readings it is easy to determine the specific gravity of milk.

1.5 The lactometer works on the Archimede's principle. This may be expressed as: "A matter cannot sink in a fluid until and unless it displaces the volume of the fluid equal to its own weight "The volume of fluid displaced depends upon the specific gravity/density of the fluid. The stem of lactometer is so calibrated that the reading is directly related to the density of milk.

1.6 The relation between the lactometer and specific gravity, density of milk can be expressed by the following formulae

1.6.1 C.L.R. = 1000 (d-1)

Where,

C.L.R. = Corrected lactometer reading

d = density of milk at 20ºC

1.6.2 Specific gravity = 1+ CLR/ 1000

2. Types of Lactometer

2.1 There are three types of lactometers which are commonly used for the determination of milk density.

2.2 Quevenne Lactometer: This is calibrated at 15.56ºC or 60ºF. It gives accurate readings in the temperature ranges of 60±10ºF subject to temperature correction factor. One can also calculate this correction factor approximately. For every 10ºf change in temperature, there is a corresponding change of 1.0 in the lactometer reading. After correction, the reading is known as Quevenne lactometer corrected reading (CLRq).

2.3 Zeal Lactometer- This is calibrated at 29ºC or 84ºF. It gives an accurate reading in the temperature range of

22ºC to 34ºC subject to temperature correction factor. If the temperature of milk is 29ºC, no correction is needed. However, for every 1ºC change in temperature, there is a corresponding change of 0.3 in the lactometer reading. After applying the correction, the reading is known as zeal lactometer corrected reading (CLRz)

2.4 ISI Lactometer: This is a new lactometer as recommended by the Indian Standards Institute. It is calibrated at 27ºC. The milk sample is kept at 27ºC and the reading of this lactometer is noted. This reading is known as CLR_I.

3. **Solid-Not-Fat-Total Solid Analysis:-** The S.N.F. and T.S. of milk have been found to correlate with the fat content and density of milk. Knowing fat percentage and density of milk, the SNF/TS can be calculated using the well known "Richmonds" formula. As the temperature of calibration of different lactometers differs from each other, different formulae are used for calculation of SNF and TS of milk.

3.1 Formulae used for calculation of SNF and TS of milk.

3.1.1 For Quevenne lactometer:

% SNF = CLRq/4 + 0.2F + 0.14

% T.S. = CLRq/4 + 1.2F + 0.14

3.1.2 For zeal lactometer:

% SNF = CLRz/4 + 0.2F + 0.50

% T.S. = CLRz/4 + 1.2F + 0.50

3.1.3 For ISI marked lactometer

% SNF = CLR_I/4 + 0.2F + 0.60

% TS = CLR_I/4 + 1.2F + 0.60

Where CLR = Corrected lactometer reading

F = % Fat in milk

4. **Precautions**

- Ensure that milk has been warmed to 40º-45ºC and cooled to the prescribed temperature before taking the lactometer reading.

- Avoid formation of air bubbles of froth
- See that the lactometer floats freely and does not touch the sides of the lactometer jar.
- Do not allow a time lag between the lactometer reading and reading the temperature of milk.

5. **Materials Required**

5.1 Milk lactometer

5.2 Thermometer (0-110ºC)

5.3 Lactometer jar

5.4 Water bath at 40º-45ºC

5.5 Tray

5.6 Beaker/bottle

5.7 Chilled water

6. **Procedure**

6.1 Determination of lactometer reading

- Heat milk sample to about 40-45ºC on water bath and maintain at this temperature for 5 minutes.
- Remove the sample from the water bath and mix the contents by rotating and inverting the bottle.
- Cool milk sample, using chilled water as near to the calibrated temperature of the lactometer as possible.
- Gently invert the sample bottle 2 to 3 times.
- Pour milk in the lactometer jar from the side of the jar (to avoid formation of air bubbles/froth).
- Pour sufficient milk into the jar so that some of it overflows when the lactometer is inserted.
- Insert the lactometer in the jar containing milk and release it at its approximate position of equilibrium.

- Do not wet more than a very short length of the stem of the lactometer above the milk surface.
- Allow the lactometer to remain steady in the jar.
- Note the scale reading of the lactometer corresponding to the upper meniscus of milk.
- Repeat the reading after pushing the lactometer about 3mm in milk and allowing it to come to rest.

6.2 Determination of per cent fat in milk.

- Estimate the per cent fat in milk as described in determination of milk fat by Gerber method.

6.3 Determination of density

- Determine the density of milk using the lactometer reading and the formula C.L.R. = 1000(d-1)

6.4 Determination of SNF and TS

- Estimate the SNF and TS content of milk from the CLR and per cent fat of milk sample using the formulae as described in 3.1

Chapter 15

Detection of Urea in Milk

1. **General:-** Urea added to milk may be detected by adding sodium hydroxide followed by sodium hypochlorite and phenol solutions to protein-free filtrate of milk, which gives the charecteristic blue/bluish-green colour for urea.

2. **Precautions-**

   - Critically note the change of colour during analysis.
   - Conduct the tests carefully using the correct procedures.

3. **Materials Required**

   3.1 Apparatus

   3.1.1 Conical flasks (50 ml)

   3.1.2 Boiling water bath

   3.1.3 Test tube

   3.1.4 Filter paper (Whatman 42 No.)

   3.2 Reagents

   3.2.1 Distilled water

   3.2.2 Sodium hypochlorite solution-2%

   3.2.3 Sodium hydroxide solution – 2% (w/v)

   3.2.4 Phenol Solution – 5% (w/v)

   3.2.5 Trichloroacetic acid solution (T.CA)-24% (w/v)

   3.2.6 Buffer (Sodium acetate-acetic acid buffer): Mix equal volume of IN Sodium acetate and IN acetic acid (pH-4.75).

4. **Procedure-**

   - Take 5ml of milk in a 50ml conical flask.

- Add 5ml of sodium acetate acetic acid buffer or 24% TCA solution.
- Heat the contents for 3 minutes. No heating required in the case of TCA.
- Filter the contents through Whatman No. 42 filter paper.
- Collect 1 ml of filtrate in a test tube.
- Add to this 1 ml of sodium hydroxide solution 0.5 ml sodium hypochlorite solution.
- Mix thoroughly.
- Add 0.5 ml of phenol solution.
- Observe for the change in colour.

5. **Observation:** Development of blue or bluish-green colour
6. **Conclusion:** Presence of urea in milk.

Chapter 16

Determination of Fat in Milk by Gerber Method

1. Milk

1.1 Apparatus: The apparatus conforming to the provisions of IS:1223 (Part-1) -1970, IS: 1223 (Part-II)-1973, and IS: 1223 (Part-III)-1977 shall be used.

1.1.1 Butyrometers: 6 per cent, 8 per cent and 10 per cent scales for estimating fat in whole milk and evaporated (unsweetened) milk, double toned milk, toned milk and homogenised.

1.1.2 Butyrometers: 1 per cent and 4 per cent scale-for estimating fat in separated milk, skim milk and butter milk.

1.1.3 10-ml pipette or automatic measure – for sulphuric acid.

1.1.4 10.75 ml pipette for milk

1.1.5 1 ml pipette or automatic measure – for amyl alcohol

1.1.6 Stoppers for butyrometers

1.1.7 Centrifuge

1.1.8 Water bath.

1.2 Reagents

1.2.1 Sulphuric acid- sulphuric acid shall have a density of 1.807 to 1.812/g/ml at 27ºC corresponding to a concentration of sulphuric acid from 90 to 91 per cent by mass.

(a) The sulphuric acid shall be colourless or not darker than pale amber in colour

(b) When diluted with distilled water to a density of 1.4g/ml not more than a very slight turbidity shall occur.

1.2.2 **Amyl alcohol-** The amyl alcohol shall conform to Grade 1 of IS: 360:1964.

1.3 Procedure:

1.3.1 Preparation of butyrometers - Mark the number of the sample to be tested legibly on the bulb of the butyrometer.

1.3.2 Transfer 10 ml of sulphuric acid in to the butyrometer by means of the 10-ml pipette or the automatic measure for sulphuric acid taking care not to wet the neck of the bytyrometer with the sulphuric acid.

1.3.3 Mixing of preparation of Sample-

If the sample is fresh, warm it to approximately 27ºC and mix thoroughly but do not shake it so vigorously as to cause underfrothing or churning of the fat. Pour the sample into another clean dry vessel and back to the original. Repeat this process of pouring back and forth until a homogeneous mixture is obtained. Allow the sample to stand for three or four minutes after mixing to allow air bubbles to rise, invert the sample bottle three or four times immediately prior to taking milk for the test.

Note-1. If the sample has aged and there is difficulty in dispersing the cream layer by the above method, warming to 30ºC may be necessary for adequate mixing.

Note-2 If the sample shows evidence of slight churning, shown by the presence of white flakes, it should be slowly warmed to 34ºC to 40ºC before mixing as described above. If after this treatment a sample appears not to be homogeneous, it shall be rejected.

1.3.4 **Addition of the sample-** Measure 10.75ml of sample into the required butyrometer by means of the 10.75 ml pipette, the temperature of sample should be brought to approximately 27ºC when it is measured.

(a) The procedure to be followed in using the pipette for measuring sample into the butyrometer shall be as follows: Dip the tip of the pipette in the well-mixed sample contained in the bottle and suck in the sample until the sample rises to a short distance above the graduation mark. Close the upper end of the pipette and withdraw it from the sample. Wipe the outside of the delivery tube of the pipette, dry with a clean piece of filter paper, hold the pipette vertically and run out the milk until the top of the milk meniscus, not the bottom of meniscus, which is difficult to see is on the graduation mark. When this is achieved, insert the jet of the pipette into the neck of the butyrometer, holding the butyrometer vertically. Touch the tip of the jet to the base of the neck of the butyrometer and slant the pipette so that the delivery tube of the pipette rests on the top neck. In this position, the vertical axis of the pipette makes an angle of 45ºC with the vertical axis of the butyrometer. Holding the pipette in this position, release the finger from the other end of the pipette directing the flow of milk against the wall of the body of the butyrometer. When emptying the pipette, take care to have a gentle flow of the milk onto the surface of the sulphuric acid, preventing, as far as possible, the mixing of the two liquids. When the outflow has ceased, wait for 3 seconds, raise the pipette and then gently touch the jet of the pipette once against the base of the neck of the butyrometer and then remove the pipette. At any stage of transferring the milk, take care not to wet the neck of the butyrometer with milk.

(b) If the same pipette is used, take care to rinse the pipette with a portion of the next sample to be analyzed. Take care to measure the sample always in the correct sequence.

1.3.5 Addition of amyl alcohol- Measure 1 ml of amyl alcohol into the butyrometer by means of 1-ml pipette or the automatic measure for amyl alcohol. Do not wet the neck of the butyrometer with alcohol.

1.3.6 Insertion of stopper- Close the neck of the butyrometer firmly with the stopper without disturbing the contents. When a double ended stopper is used, screw it in until the widest part is at least level with the top of the neck. When a lock stopper is used insert it until the rim is in contact with the neck of the butyrometer.

1.3.7 Mixing of contents- Shake the butyrometer carefully without inverting it until the contents are thoroughly mixed, the curd is dissolved and no white particles are seen in the liquid. Then invert the butyrometer a few times to mix the contents thoroughly.

Note- When large number of samples is to be mixed, shake the butyrometers in a protected stand until the contents are thoroughly mixed and no white particles are seen. Invert once or twice during the process.

1.3.8 Temperature adjustment- Transfer the butyrometer quickly, with the bulb uppermost, into a water bath having a temperture of 65 $\pm$ 2ºC and leave it there for not less than 5 minutes. Take care to have the water level in the bath above the top of the fat column in the butyrometer. Meanwhile, adjust the stopper so that the fat column shall be on the scale after centrifuging.

1.3.9 **Centrifuging:-** Take the butyrometer out of the water, dry it with a cloth and transfer it to the centrifuge, placing two butyrometer diametrically opposite so as to balance the rotating disc. Centrifuge at the maximum speed for 4 minutes. Bring the centrifuge to stop gradually. Transfer the butyrometers stoppers downwards into a waterbath having a temperature of 65 $\pm$ 2ºC and allow the butyrometers to stand in the water bath for not less than 3 minutes and not more than 10 minutes.

1.3.10 **Reading of butyrometer:-** Before taking down the reading, adjust the position of the fat column to bring the lower end of the column on to a main graduation mark. When double-ended stoppers are used, do this by slightly withdrawing the stopper and not by forcing it further into the neck. Note the scale readings corresponding to the lowest point of the fat meniscus and the surface of separation of the fat and acid, the difference between the two readings, gives the percentage by mass of fat in the milk. When readings are being taken, hold the butyrometer with the graduated portion vertical, keep the point being read in level with the eye, and then read the butyrometer to the nearest half of the smallest scale division.

1.3.11 **Procedure for skim milk, separated milk, butter milk, and evaporated (unsweetened milk)** -

When these are being tested, repeat the temperature adjustment and centrifuging before take down the reading as prescribed above.

Note:- If there is insufficient fat in the butyrometer to enable the fat content to be read, record the apparent fat content, for example, nil, trace, or fraction of miniscus.

1.3.12 Procedure for homogenized milk-

In case of homogenized milk, obtain the second value of fat content. If the second value does not exceed the first value by more than half the smallest scale division of the butyrometer, the second value shall be recorded as the fat content of the milk.

(a) If the second value exceeds, the first value by more than half the smallest scale division, repeat the procedure and obtain a third value for the fat content. If the third value does not exceed the second value by more than half the smallest scale division, the third value shall be recorded as the fat content of the milk.

(b) If the third value exceeds the second value by more than half the smallest scale division, repeat the procedure and obtain fourth value for the fat content. The fourth value shall be recorded as the fat content of the milk, but if this value exceeds the third value by more than half the smallest scale division, it should be regarded as of doubtful accuracy.

Note-1 If even after several centrifugings, the fat is turbid or dark in colour or if there is white or black material at the bottom of the fat column the value for fat content would not be accurate.

Note:2 The results obtained may be slightly high.

1.3.13 Procedure for milk containing preservatives:-

(a) If the milk containing preservatives has gone through the process of homogenization, follow the procedure described in 1.3.12. In case the milk containing preservatives is skim milk, follow the procedure described in 1.3.11.

(b) In milk containing preservatives, there may be some difficulty in achieving complete solution of the protein. In such cases, place the butyrometer, stopper downwards in the waterbath maintained at 65 $\pm$ 2º with occasionally shaking and the inversion of the butyrometer until no white particles are seen. Then proceed as described in 1.3.8, 1.3.9 and 1.3.10

Note:- If the time required in the waterbath to dissolve the protein exceeds 10 minutes, the method would not give an accurate result and would not be applicable to the sample.

1.3.14 Precautions:- If a fluffy layer is observed at the base of the fat column in the butyrometer, reject the test. Examine the stopper to see if it is in good condition, repeat the test and take greater care to ensure that the curd is completely dissolved.

(a) If the fat column is so dark as to make reading difficult, reject the test and check the strength of the sulphuric acid.

(b) If a large number of sample has to be tested, it is preferable to use automatic measures for measuring the sulphuric acid and amyl alcohol, especially the latter, otherwise there is a possibility of injurious effects to the health arising from the inhalation of amyl alcohol vapours by the use of 1ml pipette.

1.3.15 Repeatability- The difference between the results of two determinations carried out simultaneously, or in rapid succession, by the same analyst shall not exceed the value corresponding to one smallest scale division of the butyrometers. When butyrometers with scale errors less than 0.01 per cent are used, the difference between the results of two determinations obtained shall not exceed the value corresponding to half the smallest scale division of the butyrometer.

1.3.16 Test report- The test report shall show the method used and the result obtained, including the following:-

(a) The capacity of the milk pipette,

(b) The scale range of the butyrometer, and

(c)· Any observation that indicates that the result is of doubtful accuracy.

Chapter 17

Determination of Milk Fat by Electronic Milk Tester

1. **Introduction-**

1.1 Basic objective in the Dairy Development programme is to increase the production of milk. Milk collection depends on prompt payment of fair and correct price to encourage producers to increase milk production. The system of payment based on quality (i.e. fat content) discourages adulteration and encourages producers to increase production of high quality milk. It is further very necessary that each sample of milk should be tested for its quality and testing should be completed within 2 to 3 hours time in order to make correct payment. Also milk sample would get spoiled if not tested immediately in the absence of facility to preserve them, specially at village level.

1.2 Age-old traditional 'GERBER' method of testing by chemicals has many inherent drawbacks, such as, human error, multi-step method, handling of corrosive chemicals and different type of glassware. All these add to the cost and time of milk testing. A quicker, reliable and economical method of milk fat testing has therefore become inevitable and immediate problem to solve. In the light of some of the problems faced by 'GERBER' method of testing, it was felt prudent, to evolve a system which should solve these problems.

1.3 Electronic Milk Tester (EMT) is simple and economical but an accurate milk fat testing instrument. It measures the fat content instantaneously on digital read-out. It does not involve the use of corrosive chemical. It works on light scattering principle with manual homogenization. It operates on AC-mains as well as on battery with inbuilt battery charger and automatic switch-over to battery in case of power failure.

1.4 The following aspects on Electronic Milk Tester are highlighted;

(a) Basic principle involved

(b) Diluent preparation

(c) Key to controls and connections

(d) Installation

(e) Operation instructions.

The calibration and fault location have been very clearly described. The useful hints on operation and preventive maintenance have been specifically detailed to benefit the users.

2. Technical Specifications

Measuring range	- 0-13% fat		
Capacity	120-150 Samples per hour.		
Accuracy (Sd)	0-5% fat	:	0.06%
	5-8% fat	:	0.10%
	8-13% fat	:	0.20%
Repeatability (Sd)	0-5% fat	:	0.03%
	5-8% fat	:	0.04%
	8-13% fat	:	0.08%
Sample Volume	0.5 ml/test		
Diluent Volume	6.5 ml/ test		
Calibration	Two individual calibration, channels, each adjustable independently within the range 0-13%		
Power supply	AC: 220/240V, Maximum + 10% Minimum - 15%		

	DC: 12V, Motor Car battery. A fully charged battery will last for at least 10 hours operation.
Ambient temperature	5-45ºC
Dimension (HxWxD)	23x31x53 cm.
Weight (without diluent)	16 Kg.

3. Principle of measurement:

3.1 Electronic milk tester is based on the photometric measurement of light scattered by the milk sample. The light is scattered by the fat globules, acting as a small prism.

3.2 Not only fat globules in the milk contribute to the light scattering, but also the proteins may effect the measurement. To eliminate their influence, it is necessary to dissolve them. EDTA solution is used for this purpose.

3.3 All the fat globules do not have the same size. The measuring system requires a constant globule size to provide a fixed relation between the amount of light scattered to the fat content. The range of globule size is limited in the milk tester to a very narrow region outside the natural range. This is achieved by homogenizing and bringing the globule size into the range of 0.5 to 1.5 microns.

3.4 Light rays from a photo lamp pass through the layer of fluid in the cuvette and are scattered according to the amount of fat globules in the sample. More the fat present in the cuvette, more is the light scattered and less light passes through the cuvette. The rays that do pass through the cuvette hit a photocell, producing a current proportional to the light intensity. The current is fed to a digital readout unit which gives direct fat percentage readout.

3.5 For better understanding of principle of measurement photometer and its associated parts are described below:-

3.5.1 **Photometer:-** The photometer consists of a lamp, lamp housing and detector assembly. The detector assembly includes the cuvette and photocell.

3.5.2 **Cuvette:-** It is made of two hardened glass discs. One has a ground depression of 0.4 mm the other has two small holes to provide inlet and outlet to the cuvette.

3.5.3 **Photocell:-** It is a selenium barrier/silicon detector photocell. Care should be taken not to expose this photocell to light when not in operation

3.5.4 **Lamp:** 12 volts tungsten lamp.

4. Identification of Controls and Connections

4.1 Front View

1. Milk-in button
2. Mix-out button
3. Meas switch (for choosing calibration and for electronic check)
4. Repeat button
5. Adjust screw, calibration -I
6. Adjust screw, calibration –II
7. Curve screw, calibration –I
8. Curve screw, calibration-II

(Items 5–8: Covered by a metal plate.)

9. Zero setting knob
10. Milk intake tube
11. Mix intake tube
12. Intake valve
13. Screw for bleeding pump
14. Diluent syringe window
15. Milk syringe window
16. Pump handle

17. Display

4.2 Rear View

18. Mains on/off switch (does not switch off battery)
19. Mains socket
20. Inlet stub for diluent
21. Outlet stub for waste
22. Main fuse (1A for 240V)
23. Battery fuse (10A)
24. Mains voltage selector (240V AC)
25. 12V battery socket
26. Switch for selecting type of power supply (lineorline/ battery) can also switch off power supply from battery to EMT (charge).

5. **Preparation of diluent:-**

5.1 Diluent is used to dilute the milk sample and dissolve the proteins.

5.2 Chemicals required to prepare 10 litres of diluent are as follows:-

(i)	EDTA sachet	-	52.6 gms. containing EDTA Powder= 45.0 gms. Di-Sodium hydroxide = 7.6gm
(ii)	Triton-X-100 (emulsifier)		0.5ml (10 drops)
(iii)	Anti-foam		0.5ml (10 drops)

5.3 Procedure:- Take a clean 10 litre plastic container and add one litre clean water. Add contains of EDTA sachet, containing diluent powder for 10 litres solution. Add 0.5ml (10 drops) of Triton-x-100 and 0.5ml (10 drops) of Anti-foam. Put the lid on the container and shake it until all the chemicals are dissolved. Then add clean water until the contents are full 10 litres and shake again to mix the solution.

5.4 The pH of this solution should be between 9.5 to 10.1. Diluent will remain good for 2 weeks before the anti-foam becomes inactive. Diluent can be kept longer if anti-foam is added to the solution from time to time (0.5ml for every 2 weeks).

6. Installation:-

6.1 Place of installation should be such that there should be sufficient space for diluent container and battery also. Pump handle of EMT must be easy to reach.

6.2 Check 1 Amp. fuse in the mains fuse holder and 10 Amp. fuse in the battery fuse holder.

6.3 Connect power cable on mains.

6.4 Use only 12V motor car battery. Connect red lead of battery cable to '+' pole of battery and black lead to '- ' pole. Connect other end of the cable to electronic milk tester.

6.5 Set selector switch at line/battery and mains switch at on. Power will now be supplied to EMT from battery. As long as mains switch is at On position, current from the mains will keep the battery charged. If the power fails, battery will still be able to supply enough current for several hours of operation.

Set the selector switch at line, if for any reason, the EMT is to be operated on mains current without a battery.

6.6 Prepare the diluent in the container as described earlier in 5.

6.7 Connect one end of the plastic pipe to the INLET stub at the back of EMT and attach the filter unit to the other end which is then placed in the diluent container.

6.8 Connect one end of the plastic pipe to the OUTLET stub and place the other end in the waste container.

6.9 Place empty beaker under milk intake tube and push "MILK IN" and "MIX OUT" buttons alternately until no bubbles are seen in syringes. End by pushing 'MIX OUT' button.

7. Preparation of Milk Samples:

7.1 **Sampling:-** The samples measured should be taken from milk in good condition, i.e. the milk should not have started to curdle or separate, and it should be free of dirt. Since milk fat is of lower density than the other milk constituents it tends to rise to the surface. Gently stir the bulk milk just before sampling to make sure that sample is a truly representative of the bulk. Fresh unpreserved samples must be tested immediately. Just before testing, turn the sample upside down a few times to mix it properly.

7.2 **Preservation:-** If measurement cannot be made immediately after sampling, then samples can be preserved for maximum 12 hours, without refrigeration, by adding 1ml saturated potassium dichromate solution in 100 ml milk. If they are to be kept longer for measurement and if transportation is necessary, then cool the preserved sample to 5-10ºC but never freeze them. Samples which have been cooled will be easy to mix if they are warmed up to 30-40ºC before being measured.

8. Operation-

8.1 **Warm-up:-** If the instrument has been switched off, then it must be allowed to warm-up before starting measurement. This is done by switching on the EMT half hour to one hour before measurement.

8.2 **De-airing-syringes:-**

The flow system must be free of air bubbles to ensure proper measurement. Presence of any air bubble in the syringe must be removed as follows:

Place empty mix beaker under milk in-take tube and push 'MILK IN and MIX OUT buttons alternately until no air bubbles are seen in the syringes. End by pushing 'MIX OUT' button.

8.3 **Zero check:-** The zero setting knob is used to adjust the read out to 0.05 when the cuvette contains pure diluent.

Zero check described below should be carried out daily before starting measurement and then once every half hour for the first 2 hours and then once an hour.

Press REPEAT button to get second decimal on the display. Place the clean, empty mix beaker under the milk in-take tube and press "MILK IN" and 'MIX OUT' button twice alternately to fill the mix beaker with diluent. Move the mix beaker to the mix intake tube and operate the handle six times up and down.

Turn the Zero knob to adjust the readout to 0.05. After Zero setting, press the REPEAT button to get one decimal readout.

Please note that zero knob when turned anti-clockwise will not display below 0.01 in series 606-001 and further.

8.4 **Measurement:** After de-airing and Zerocheck, measurement can begin. First result after zerocheck/zero setting should not be recorded as it will be little low. Therefore, measure the first sample after zero check twice and record only second result.

Turn the milk sample gently upside down a few times. Place the sample under milk in-take and fully press 'MILK IN' button.

Slowly remove the sample without touching the milk in-take tube. Place clean mix beaker under milk in-take so that it fits into the notch on the side of the EMT. Press 'MIX OUT' botton in all the way to dispense milk and diluent into the mix beaker. Move the mix beaker to the mix in-take tube and position it so, that it rests in the notch. Operate the homogenizer handle up and down three times in a steady measurement. When the handle is pressed down the third time, let it rest in bottom position and result will soon appear on the display. Empty the mix beaker completely and it is ready for the next sample.

8.5 **End of Measurement:-** Place clean, empty mix beaker under milk in-take tube and press 'MILK IN' and MIX OUT' buttons twice alternately to fill mix beaker with diluent.

Place the mix beaker under mix in-take and operate the handle up and down six times to flush the cuvette. switch-off the EMT.

8.6 **De-airing the Homogenizer:-** Air will enter the homogenizer if the handle is raised when there is no liquid present at the mix in-take tube.

Place the mix beaker under milk in-take tube and press 'MILK IN and 'MIX OUT' button twice to fill the beaker with diluent. Raise the handle. Loosen the bleeder screw on the top of mix in-take value. Let the handle drop by its own weight and tighten the bleeder screw again. Operate the handle three more times up and down to be sure that all air is out of the system.

8.7 **Electronic Check:** Set 'MEAS' switch to CHECK. Display should show 5.00 $\pm$ 0.1. This is a check of digital PCB but not of lamp and photocell.

9. **Calibration:-**

9.1 The purpose of calibration is to adjust the instrument to give results over the whole measuring range. There are two steps in calibration. One to give a correct linearity and other to give correct results when compared with the results of a reference method.

9.2 If few milk samples of different fats are measured, one expects the instrument to show these results within some small limit. If all results are too high or too low by the same proportion the linearity is all right. Then test the known high fat reference sample on EMT and turn appropriate 'Adjust' screw until correct value appears on display. If high fat results are too low and low fat results are too high or vice versa, then the linearity must be adjusted as follows:-

Take one sample of high fat content (Actual fat is of no importance)

Linearity:-

(i) Results of Half Value 1......................................

2. ....

3. ....

4. ....

A Average ....

(ii) Results of Full Value

1. ....

2. ....

3. ....

4. ....

B Average ....

(iii) B - 2 A = C

(iv) CX -3 = D

(v) From B adjust with CURVE screw up or down by D, then adjust with ADJUST screw down or up to give 2A.

Full Value:- Measure the sample in a normal way to get full value of the measurement.

Half value:- Place empty mix beaker under milk in-take and press 'MILK IN' and 'MIX OUT' button to fill the mix beaker (i) Place milk sample under milk in-take tube and press 'MILK IN' button. Place mix beaker under milk in-take tube and press 'MIX OUT' button to dispense milk and diluent into the mix beaker.

(ii) Remove this mix beaker and place mix beaker (i) Containing only diluent, under milk in-take tube and press 'MILK IN' button and remove this mix beaker. Place the mix beaker (ii) Once more under milk in-take tube and press 'MIX OUT' button. This will give double volume of diluent, so that result should be half of that obtained in the normal way.

9.3 An Example will simplify the procedure of linearity adjustment.

9.3.1 The average of half value = 4.61

The average of full value = 9.44

9.44 - (2x4.61) = 9.44 - 9.22 = + 0.22

This means the high results are more than twice the low results. Multiply the difference by 3-(in this case –3 x 0.22 = - 0.66). With the small screw driver, turn the CURVE screw from 9.44 to 8.78 (9.44-0.66 = 8.78).

Then turn the corresponding 'ADJUST' screw from 8.78 to 9.22 to get the correct value.

9.3.2 If the high results are 'low, say 9.13, the comparison will give a negative number i.e. 9.13–(2x4.61) = 9.13-9.22 = -0.09

Multiply the difference by –3 (in this case, -3 x - 0.09 = + 0.27). Which means that CURVE must be adjusted upwards instead of downwards and ADJUST must be adjusted downwards to the correct value.

9.3.3 Continue the above procedure until the difference 'C' is less than 0.04.

10. Useful hints on operation of EMT.

10.1 The instrument should be kept clean.

10.2 Half hour to one hour warm-up, de-airing syringes and zero check must not be forgotten before starting the measurement.

10.3 When pressing the 'MILK IN' and 'MIX OUT' buttons. Use the ridge on the cabinet to support the fingers, and press the buttons in evenly with the thumb. This will improve your instrument's performance.

10.4 Chemicals must be stored in a clean and dry place.

10.5 It is not necessary to clean the milk in-take tube and mix in-take tube between each sample. If the milk in-take tube reaches far into the sample, it may be necessary to wipe it between sampling and dispensing to avoid carry-over. This happens if milk from one sample is left on the out side of the milk in-take tube and is then mixed with the next sample.

10.6 Place milk sample under the milk in-take tube so that the top of the tube is approximately 5mm below the surface. Though 5mm is not so critical but the tube should be dipped at the same distance below the surface in every sample.

10.7 During storms with lighting, unplug the EMT from mains supply to avoid danger of damage to the instrument.

10.8 EMT should preferably be connected with battery and normal position of operation is, Mains at On and selector switch at line/battery. If mains fails, battery will still be able to supply enough current for several hours of operation. If battery reserve is low then set selector switch at CHARGE, with mains at On. This switches off power supply to measuring circuit. A completely drained battery will be full charged in 8-12 hours.

10.9 Battery fuse will blow if leads are connected to wrong poles on the battery. Always remember or connect red lead of battery cable to '+ve" pole of 12V car battery and black lead to "-ve" pole.

10.10Check the acid level of the battery once a fortnight and refill it if necessary with distilled water until plates are just covered.

11. Useful hints on preventive maintenance of EMT:-

11.1 Electronic milk tester will give accurate results if:-

(i) It shows a stable zero at 0.05 ($\pm$ 0.03) up to an hour at the time during milk testing.

(ii) It shows a stable readout ($\pm$ 0.03) after the third stroke when operated on second decimal.

(iii) It shows essentially the same readout when the same sample is tested again and again.

(iv) It has been calibrated by comparing its results with results of 'GERBER' method.

However it may give misleading result if one or more of the above features fails. Therefore make a quick check every day at start-up, and a complete check once a week.

11.2 The hints below will help you to locate and remedy troubles, should they appear.

11.2.1 Display Fails:- If no, display, the supply voltage from battery or mains probably has failed or the fuse has burnt out. Use 10A fuse for battery and 1A fuse for mains. 10A fuse is intended to burn out if the battery's negative pole is connected to the EMT's positive terminal and vice versa.

11.2.2 Zero setting Fails:-

(i) First apply a double measure of diluent to the mix in-take and operate the handle six times.

(ii) If the reading is 8888 which remains unaffected by 'ZERO' control but 5.00 CHECK is Ok, the photolamp is probably burnt out. However, first check the supply voltage. Battery should be min 10.5V. Mains minimum 190V. Check loose or broken connection to lamp or photocell.

(iii) If above procedure fails to locate error or in case the reading shows Zero and cannot be raised, then either photocell or lamp or power PCB is faulty.

11.2.3 Zero Drifts:-

(i) The Zero drift means that 0.05 readout is not stable. Zero drift is normal during warm-up period because thermoblock temperature needs time to stablize.

(ii) If zero continues to drift or fluctuate or 0.05 can not be obtained then check supply voltage.

(iii) Zero drift may be caused by air in-take. Therefore, make sure that homogenizer pump is perfectly de-aired and that the mix in-take tube is firmly fixed in the lowest possible position in the slot, otherwise the volume supplied by the syringes will not be sufficient to give an air-free charge at the third pump stroke.

(iv) Even if air in-take is excluded, air bubble may appear in the cuvette (causing the readout to drift slowly upscale after pump operation and goes back very slowly) if the value V3 after cuvette outlet is leaking. A quick check for this is to raise the external waste tube to a position above the EMT to increase cuvette pressure. If this stabilizes the readout, then clean valve V3.

(v) If Zero drifts quickly up-scale by about 0.3-1.0% after pump operation and goes back slowly, then this may be due to the moisture or condensation on the outside of the cuvette glass, which must be carefully wiped dry.

(vi) In rare cases, malfunctions of the photocell or lamp may cause zero drift.

11.2.4 Readout Drifts After a Milk Test:- As low drift down-scale by 0.01-0.04% is considered normal and does not effect results on one decimal readout. If drift is more, then that is because of possible presence of moisture or leakage in the cuvette. Therefore cuvette may need new gaskets. It is important that the inside of the photometer must be carefully wiped and dry before assembly.

11.2.5 Repeatability Fails: When the same sample is tested again and again it must show repeatability within the prescribed tolerance limit. If higher variations appear, check the following:

11.2.6 Is the EMT-being operated correctly?

(i) 'MILK IN' and 'MIX OUT' buttons should be pressed smoothly and uniformly.

(ii) Don't push 'Mix-out' button, so fast that mix splashes out of the mix beaker during the discharge.

(iii) Never touch the milk in-take tube when removing the sample bottle.

11.2.7 Is the test sample in good condition?

If it is acidified, churned out, separated or otherwise non-uniform, good repeatability cannot be obtained.

11.2.8 Instrumental Checks:

(i) If at CHECK, 5.00 + 0.010 reading is stable, then digitalizer PCB is OK, otherwise it is faulty.

(ii) Operate the EMT with diluent alone (no milk). Note that there is no zero drift.

(iii) If zero is stable, the lamp cuvette and photocell are in good order and the trouble must be in the syringe assembly. The possible cause is leaking of valves V1, V2, V3 which will cause high and fluctuating results. Clean the valves.

11.2.9 Miscellaneous Problems:-

(i) If mix is not sucked in, from mix in-take tube, then clean the ball of mix in-take valve.

(ii) If diluent leaks from milk in-take tube, then clean valve V2

(iii) If fluid leaks from bleeder screw of mix in-take valve, then replace gasket of bleeder screw.

(iv) If fluid leaks from mix in-take valve when handle is pressed down, then replace the O-ring of intake valve.

(v) If 8888 or 00.00 or any random number appears on CHECK and in measuring position then digitalizer PCB is faulty.

Reference:- Rajasthan Electronics and instruments Ltd. Plot No.2, Kanakpura Industrial Area. Jaipur, India: Technical Manual on Electronic Milk Tester.

Chapter 18

Detection of Preservatives in Milk

1. **General:-** Milk and other dairy products are required by public health laws to be free from preservatives. With a few exceptions, any preservative added to milk is not removed in the process of treatment or manufacture. Thus, if milk containing preservatives is accepted, the treated milk or milk product when subsequently offered for sale may contain preservatives and render the seller liable to prosecution besides constituting a hazard to health. No preservative shall, therefore, be added to milk except in the case of the samples which have to be preserved for chemical examination.

   Some of the most common preservatives found in milk are boric acid and borax, benzoic acid formaldehyde and salicylic acid. Hypochlorite residues may also be found in milk when chlorine sterilizes are used for sterilizing milk handling equipment unless such equipment is properly rinsed with water. Anti-biotics used in the treatment of udder diseases of milch animals may also find their way in milk, if proper precautions are not taken. The presence of chlorine of anti-biotics in milk not only enhances the keeping quality of the milk but will reduce the plate counts and also lengthen the MBR or resazurin in reduction time.

   When testing for preservatives, it is necessary to carry out a control test with a milk sample known to be free from any preservative.

2. **Turmeric Paper Test for Boric Acid or Borax:-**

   2.1 Reagents:-

   (a) Turmeric paper - dried

   (b) Concentrated hydrochloric acid-sp. gr. 1.16.

   (c) Ammonium hydroxide-sp gr. 0.88

(d) Lime water or caustic soda solution.

2.2 Procedure:- Immerse a strip of turmeric paper in a sample of milk previously acidified with hydrochloric acid in the proportion of 7 ml of concentrated hydrochloric acid to each 100ml of milk. Allow the paper to dry spontaneously. If boric acid or borax is present, the paper will acquire a characteristic red colour. The addition of ammonium hydroxide will change the colour of the paper to a dark green, but the red colour may be restored by hydrochloric acid.

Alternatively, make about 25 ml of the sample strongly alkaline with lime water or caustic soda and evaporate to dryness on a waterbath. Ignite the residue at a low red heat to destroy organic matter, cool, digest with about 15ml of water, add concentrated hydrochloric acid, drop by drop, until the ignited residue is dissolved. Then add 1 ml in excess. Saturate a piece of turmeric paper with this solution and allow the paper to dry without heating. The colour change will be the same as described above.

3. Hehner Test for Formaldehyde:-

3.1 Reagent:- Concentrated sulphuric acid-commercial; sp. gr. 1.84.

3.2 Procedure:- To about 10ml of milk in a wide-mouthed test tube add about half the volume of concentrated sulphuric acid pouring the acid carefully down the side of the tube so that it forms a layer at the bottom without mixing with the milk. A violet or blue colour at the junction of the two liquids indicates the presence of formaldehyde. The test is sensitive to one part in 10,000.

Note:- The test is given only in the presence of a trace of ferric chloride or other oxidizing agents. This test may be combined with the determination of fat, noting whether a violet colour forms on addition of sulphuric acid in the butyrometer.

4. Test for Benzoic Acid:-

4.1 Reagents:

(a) Dilute hydrochloric acid - 1:3 by volume.

(b) Ethyl ether

(c) Ammonium hydroxide - sp gr. 0.88

(d) Ferric chloride solution - 0.5 per cent (w/v), neutral.

(e) Sodium hydroxide solution - 10 per cent (w/v).

(f) Potassium nitrate - Crystals.

(g) Concentrated sulphuric acid - sp. gr. 1.84.

(h) Ammonium sulphide — freshly prepared and colourless.

4.2 Procedure:- Acidify 100 ml of milk with 5 ml of the hydrochloric acid. Shake until curdled. Filter and extract the filtrate with 50 to 100 ml of ether. Wash the ether extract layer with two 5 ml portions of water. Evaporate the greater portion of ether in a porcelain dish on a waterbath and allow the remainder to evaporate spontaneously. If benzoic acid is present in considerable quantity, it will crystalize from the ether in shining leaflets and give a characteristic odour on heating.

Dissolve the residue in hot water, divide into two portions and test as follows:-

(a) Make one portion alkaline with a few drops of ammonium hydroxide, expel the excess of ammonia by evaporation, dissolve the residue in a few millilitres of hot water, filter, if necessary. Then add a few drops of the neutral ferric chloride solution. A salmon coloured precipitate of ferric benzoate indicates the presence of benzoic acid.

(b) To the other portion add one or two drops of sodium hydroxide solution and evaporate to dryness. To the residue, add five to ten drops of concentrated sulphuric acid and a small crystal of potassium nitrate. Heat for 10 minutes in a glycerol bath at 120 to 130ºC or for 20 minutes in a boiling waterbath. The temperature of glycerol bath shall not exceed 130ºC. After cooling, add 1 ml of water, make distinctly ammoniacal and boil the solution to decompose any ammonium nitrate that may have been formed, cool, pour into a test tube, and add a drop of fresh colourless ammonium sulphide without allowing layers to mix. A red-brown ring indicates benzoic acid. On mixing, the colour diffuses through the whole liquid, and on heating finally, changes to greenish yellow. This differentiates benzoic acid from salicylic acid. The latter forms coloured compounds which are not destroyed by heating.

5. **Ferric Chloride Test For Salicylic Acid:-**

5.1 **Reagents:-** (a) Dilute hydrochloric acid – 1:3 by volume.

(b) Ethyl ether.

(c) Ferric chloride-0.5 per cent, neutral.

5.2 **Procedure:-** Acidify 100ml of milk with 5ml of dilute hydrochloric acid. Shake until curdled and filter. Extract with 50 to 100 ml of ether. Wash the ether layer with two 5 ml portions of water. Evaporate the greater portion of ether in a porcelain dish on a steam bath, allow the remainder to evaporate off. Add one drop of the ferric chloride solution. A violet colour indicates the presence of salicylic acid.

6. **Test for Hydrogen Peroxide:-**

6.1 **Reagent:-** Paraphenylenediamine solution- 2 per cent (w/v)

6.2 **Procedure:-** Add to about 5 ml of milk in a test tube, an equal volume of raw milk, followed by five drops of a 2 per cent solution of paraphenylenediamine. A blue colour is developed in the presence of hydrogen peroxide.

Note:- Hydrogen peroxide is destroyed when milk is heated or stored for a long interval.

7. **Detection of Hypochlorite**

7.1 **Apparatus:-** Apparatus

(a) Centrifuge.

(b) Tubes for Centrifuge – each of capacity 12.5ml.

(c) Mercury vapour lamp – fitted with a wood's filter.

7.2 Reagent:-

Stannous chloride solution: 0.025 percent (w/v) in 73.5 percent sulphuric acid (prepared by mixing three volumes of concentrated sulphuric acid and one volume of water).

7.3 Procedure:- Cool about 3 ml of milk, taken in a test tube in a freezing mixture of ice and salt to 2 to 5ºC. In another test tube take an equal volume of the stannous chloride solution and similarly cool and add to milk. Shake the tube whilst in the freezing mixture and hold for 3 minutes. Place the mixture in a 12.5 ml centrifuge tube and centrifuge for 3 minutes at 2500 rev/min. A yellow-green colour is produced in the presence of hypochlorite. Alternatively, after centrifuging, examine the tube in ultraviolet light from a mercury vapour lamp fitted with a woods filter for the presence of any yellow fluorescence.

Chapter 19

Detection of Neutralizers in Milk

1. **General:-** Neutralizers in the form of lime water or sodium bicarbonate may be added to neutralize developed acidity before milk is processed. Such a practice is not permissible.

2. **Rosalic Acid Test for Carbonates:-** To about 5 ml of milk in a test tube, add 5 ml of alcohol, a few drops of a one per cent (w/v) alcoholic solution of rosalic acid and mix. If a carbonate is present, a rose red colour appears whereas pure milk shows only a brownish colouration.

3. **Test for Alkalinity of Ash:-** Neutralization of milk, whether with lime, soda ash or caustic soda invariably increases the ash content and total alkalinity of the ash from a fixed quantity of milk. This is detected by ashing accurately measured 20 ml of milk and titrating the ash after dispersing in 10 ml of water. The amount of standard 0.1 N hydrochloric acid required to neutralize the alkalinity shall not exceed 1.20ml.

Chapter 20

Detection of Adulterants in Milk

1. **General:-** The modes of adulteration commonly encountered in market samples are:-

   (a) removal of fat by skimming,

   (b) addition of separated milk or skimmilk to whole milk

   (c) addition of water, and

   (d) addition of starch and cane sugar for raising density to prevent detection of added water by lactometers.

2. **Detection of Skimming:-** An indication of the removal of excess fat from milk is given by the following:-

   (a) Lower percentage of fat,

   (b) higher density reading of the sample at 27ºC and

   (c) higher ratio of solids-not-fat: fat

3. **Detection of Milk mixed with Separated Milk or Skimmilk:-**

   When fresh separated milk or skim milk has been added to whole milk, it could be inferred from the following facts:

   (a) Lower percentage of fat.

   (b) Higher density of the toned milk sample at 27ºC

   (c) Higher percentage of solids-not-fat and

   (d) Higher ratio of solids-not-fat; fat

4. **Detection of Extraneous Water:-**

   Presence of extraneous water in milk is detected by the following facts:-

   (a) Lower percentage of fat.

   (b) Lower density of milk at 27ºC.

(c) Lower percentage of solid-not-fat and

(d) Depression of freezing point.

5. Detection of Starch:-

Starch or cereal flours may be added to make up the density of milk to prevent detection of added water. The presence of starch or cereal flours is detected by the following test;

Place in a test tube about 3ml of well-mixed sample. Bring it to boil by holding the tube over a flame. Allow it to cool to room temperature. Add a drop of one per cent iodine solution. Presence of starch is indicated by the appearance of a blue colour which disappears when the sample is boiled and re-appears on cooling.

6. Detection of Cane Sugar:-

Cane sugar may be added to milk to raise the density to prevent detection of extraneous water. It is detected by the following test:-

To about 15 ml of milk in a test tube, add 1 ml of concentrated hydrochloric acid and 0.1 g of resorcinol and mix. Place the tube in boiling waterbath for five minutes. In the presence of cane sugar, a red colour is produced.

Chapter 21

Determination of Aflatoxin M1 Content in Milk by High Performance Liquid Chromatography (HPLC)

1. **Principle:**

   In this method, aflatoxin M1 is extracted from the test sample with the use of chloroform. The extract obtained is purified by passing through a chromatographic column. The purified extract is evaporated to a known volume and subjected to high performance liquid chromatography (HPLC). In HPLC, determination of the amount of aflatoxin M1 is performed by comparison of the area of the Alfa toxin M1 peak from the sample extract with the area of the peaks obtained by injection of known amounts of standard aflatoxin M1 into the chromatograph.

2. **Apparatus:**

   - Rotary vacuum evaporator, with a 500ml round bottom flask.
   - Glass column for column chromatography: length 30 cm, internal diameter 1 cm, sintered disk at the bottom.
   - HPLC injection syringes of capacity 50 and 100 ml.
   - Drying oven capable of operating in the range of 75 to 105ºC.
   - Vortex mixer.
   - Auto pipette, variable volume 100 µl to 1000 µl
   - HPLC apparatus, consisting of an injector, solvent delivery system, fluorescence detector capable of operating as follows; λex 360nm and λem 420 nm.

- Analytical (HPLC) column: Reverse phase C18 column.

3. **Reagents:-**

- Chloroform, stabilized with 0.5% of ethanol 96% (v/v)
- Toluene
- Acetic acid (>99.7%)
- Acetonitrile
- Methanol
- Acetone
- n-Hexane
- Diethyl ether (Peroxide free, with an ethanol content less than 0.05% v/v)
- Sodium dodecyl sulfate
- Sodium chloride solution, saturated: Mix approximately 400g. of sodium chloride with 1 litre of water and allow to stand overnight.
- Toluene/acetic acid mixture (9+1) (v/v)
- Acetonitrile/Diethyl ether/n-Hexane mixture (1+3+6)
- Chloroform/Acetone mixture (8+2) (v/v)
- Mobile phase for HPLC: Acetonitrile/water mixture (25+75) (v/v) (HPLC quality)
- Sodium sulfate, anhydrous, coarse grained.
- Silica gel 60 Merck, for column chromatography, particle size 63-200 μm. Dry for 1 hr at 105ºC in a porcelain dish. Transfer the silica gel to a conical flask and stopper the flask. After cooling, add 1 ml of water per 100g and restopper the flask. Shake well for 1 min and leave to settle for at least 15 hr.
- Inert gas, for example nitrogen.
- Aflatoxin M1 standard solution.

(a) Ampoules containing 2.5 ml of a standard solution of Aflatoxin M1 of concentration 10 μg/ml in chloroform (commercially available from National Institute of Public Health and Environmental Hygiene (RIVM), Netherlands) OR

(b) Ampoules containing 0.05 mg of Aflatoxin M1 (Sigma make).

4. **Aflatoxin M1 Stock Solution:**

4.1 When using an ampoule containing 2.5 ml of a standard solution of Aflatoxin M1, quantitatively transfer its contents to a 25ml volumetric flask by rinsing the ampoule with chloroform and adding the rinsing to the contents of the volumetric flask. Dilute the contents of the flask to the mark with the chloroform. This solution contains 1 µg/ml of aflatoxin M1. The flask containing the solution should be well stoppered and wrapped in aluminium foil to exclude light. When not in use, store it in the dark at a temperature below 5ºC.

4.2 When using an ampoule containing 0.05 mg of Aflatoxin M1, quantitatively transfer its contents to a 50ml volumetric flask by rinsing the ampoule with chloroform and adding the rinsing to the contents of the volumetric flask. Dilute the contents of the flask to the mark with the chloroform. This solution contains 1µg/ml of Aflatoxin M1.

5. **Working solution for HPLC:-**

5.1 By means of a pipette, transfer 0.5ml of the µg/ml aflatoxin M1 stock solution to a 10ml conical flask. Evaporate the solution to dryness using a current of inert gas and dissolve the residue obtained in 10 ml of acetonitrile/water mixture. This solutions contains 0.05 µg/ml of Aflatoxin M1. Use this solution for the preparation of a series of dilutions of Aflatoxin M1 standard solution, which contain 0.0025, 0.005, 0.01 and 0.025 µg aflatoxin M1/ml respectively.

5.2 By means of a pipette, transfer 5 ml of the µg/ml aflatoxin M1 stock solution to a 10ml conical flask. Evaporate the solution to dryness using a current of inert gas and dissolve the residue obtained in 10 ml of acetonitrile/ water mixture. This solution contains 0.5μg/ml (0.5 mg/ ml) of Aflatoxin M1.

6 **Procedure:-** As far as possible the work should be carried out with daylight excluded.

6.1 Extraction:-

Cool the milk, the sodium chloride solution and the chloroform to approximately 4ºC. Transfer 50ml of milk to a separating funnel. Add 10 ml of sodium chloride solution and swirl the separating funnel gently in order to mix the contents.

Add 125 ml of chloroform (cooled to appr. 4ºC) and to take and invert the separating funnel repeatedly for 1 minute so as to completely mix the aqueous and chloroform phases yet avoid the formation of intractable emulsions. Allow the layers to separate.

Drain off the chloroform layer (bottom layer) into a conical flask Add approximately 5g of anhydrous sodium sulfate and leave the mixture to stand for 15 minutes, shaking occasionally. Filter through fluted filter paper and collect 75ml of filtrate in a measuring cylinder (if less than 75ml of filtrate is obtained, the calculation should be adjusted).

6.2 Column Chromatography:-

Prepare a silica gel column as follows: Fill the column with 10 ml of chloroform. Remove any air bubble present. Add 1 g of sodium sulfate, rinse the inside of the column wall with approx. 1 ml of the chloroform to remove adhering sodium sulfate particles and ensure that the surface of the sodium sulfate layer is horizontal. Weigh 2 gm of silica gel, pour it into the column immediately, add 5 ml of chloroform and stir the mix. Until the silica gel is completely dispersed through the chloroform and all air is removed. Drain off the chloroform until the level is 3 cm above the silica gel layer. Once the silica gel has settled, carefully cover it with approx. 2g of sodium sulfate. Drain off the chloroform until the level is just above the sodium sulfate layer.

Transfer the 75ml of filtrate to the column and elute by gravity until the level of liquid has reached the top surface of the sodium sulfate layer. Do not allow the column to run dry.

Then elute with 25ml of toluene/acetic acid mixture (9+1) (v/v) until the level of liquid has just reached the top surface of the sodium sulfate layer. Discard the elute.

Elute with 25 ml of n-hexane until the level of liquid has just reached the top surface of the sodium sulfate layer. Discard the elute.

Elute with 25ml of acetonitrile/diethyl ether hexane until the level of liquid has just reached the top surface of the sodium sulfate layer. Discard the elute.

Finally, elute aflatoxin M1 from the column with 60 ml chloroform/acetone and collect elute in a round bottomed flask. Evaporate until dry in the rotary evaporator and quantitatively transfer the residue, using chloroform to a 4 ml conical tube.

6.3 Evaporation:-

Evaporate the above solution in a current of inert gas at a temperature of approximately 50ºC until dry. Avoid overheating the (dry) extract. When the tube has cooled down, add 100 µl. of acetonitrile water mixture using an injection syringe or auto pipette. Mix by means of a vortex mixture for 1 min.

6.4 Determination:-

Pump the eluent at a constant speed through the HPLC column, inject in sequence, 40 ml of the aflatoxin m1 working solutions (0.05 µg/ml), equivalent to 0.1, 0.2, 0.4 and 1µg of aflatoxin M1 (inject 2,4,6,8,10 µl (0.5 µg/ml) equivalent to 1,2,3,4,5 µg of aflatoxin m1).

Prepare a calibration graph by plotting the peak areas for each standard as prepare a calibration graph by plotting the peak areas for each standard as a function of the quantity of aflatoxin M1 injected. Inject 40 µl of sample extract into the HPLC apparatus using the same conditions as for the standard solutions.

Determine the area of the aflatoxin M1 peak of the sample extract. From the calibration graph determine the amount of aflatoxin M1 in µg. If the peak area of the sample extract is greater than that of the highest standard solution, dilute the extract quantitatively with mobile phase and again inject into the HPLC apparatus.

Analytical Condition:

Mobile phase	- λ Acetonitrile/water (25:75)
Stationary phase	- C18 (ODS) column, 25 cm long, 4.6 mm dia, 5µ particle size
Flow rate	- 1ml/min
Detector	- Fluorescence detector
Wave length	- Exi; 360 nm, Emi: 420nm.

7 Calculation:-

The aflatoxin M1 content in the sample of milk, expressed in µg/lit; is given by the formula; (Maf x Vext)/((VMxVx (Vf/125)).

Where,

Maf.: mass, in ng, of aflatoxin M1 determined from the calibration graph.

Vm: volume, in microlitre, of sample extract injected.

Vext: volume, in microlitre, in which the sample extract is dissolved.

V: volume in ml of milk

Vf: volume, in ml of the filtrate obtained in the extraction stage.

125: volume, in ml, of chloroform used in the extraction.

7.1 Repeatability:

The ratio between two single results found on identical

test material by one analyst within a short time interval should not exceed 1.9 in HPLC determination.

7.2 Reproducibility:-

The ratio between two single and independent results found by two operators working at different laboratories on identical test material should not exceed 3.2 in HPLC determination.

Reference:

1DF Standard 111A: 1990

Chapter 22

Estimation of Lead in Milk by Atomic Absorption Spectrophotometer (AAS)

1. **Principle:-** Sample is dry ashed and residue is dissolved in dilute Nitric Acid. Lead content is determined by Atomic Absorption Spectrophotometer at 217 nm.

2. **Apparatus:-**
   - Atomic Absorption Spectrophotometer (AAS) Ashing vessels:
   - Ashing vessels: Approx. 100ml, flat bottom, pt or silica evaporating dishes.
   - Furnace with pyrometer to control range of 250-600ºC with variation of <=10ºC.
   - Forced draft oven

3. **Reagents:-**
   - Nitric Acid: AR grade
   - Glass double distilled water
   - Lead standard solutions
   - Stock Solution: 1.5985 gm of purified Lead Nitrate (99%) is dissolved in 1 lit of 1N Nitric acid (1 mg/ml i.e. 1000 ppm).
   - Intermediate Solutions: Pipette 10 ml stock solution into 1 lit. volumetric flask and dilute to volume with Nitric Acid (100 ppm).
   - Working solution: Pipette 2,4,6,8,10 ml intermediate solution in to 100ml volumetric flasks, and dilute to volume with IN HNO3 (i.e. 2,4,6,8,10 ppm respectively).

4. **Procedure:-**

- For milk, 50 gms of sample is taken in ashing vessel.
- Dry the sample overnight at 120ºC in oven or burn directly on gas burner or hot plate to remove all the volatile matters (sample must be absolutely dry to prevent flowing or spattering in furnaces)
- Transfer above sample in the Muffle Furnace maintained at 250ºC, then allow to increase the temperature upto 500ºC.
- Allow to keep the sample in the Muffle Furnace for 16 hrs (till complete ashing) at 500ºC or 6 to 8 hrs at 550ºC.
- Allow to cool it for overnight in the Muffle Furnace itself.
- Ash should be complete white or reddish gray and free from the black carbon particles.
- If any black particles are found, then ash should be wetted with minimum amount of water followed by drop wise addition of Nitric Acid (0.5ml to 3 ml). Dry on hot plate and transfer the ash residue in Muffle Furnace for 2 hrs. at 550ºC.
- Dissolve the final residue in 5 ml 1N Nitric Acid, warming on hot plate to aid solution. Filter if necessary, by decantation through ash less filter paper into 25ml vol. flask.
- Repeat with two 5ml portions 1N Nitric acid, filter and add washings to original filtrate. Dilute to volume with 1N Nitric acid, filter and add washings to original filtrate. Dilute to volume with 1N Nitric acid.
- Along with the sample, blank solution is prepared by following the above steps.

5. **Determination:**

- Set instrument to the following working conditions;

Wave length	:	217 nm
Slit width	:	1.0 nm.
Lamp current	:	5 m. A.
Flame	:	Air Acetylene

- Before ignition, optimize the burner position in such a way that the light coming from Hollow cathod lamp (lead lamp) focus on the target area of cleaning strip.
- Adjust the lamp position by screw to get the maximum intensity (absorption) at 217 µm.
- Ignite the flame, aspirate to working solution of lead (2,4,6,8,10 ppm) to get the calibration curve (it should be straight line)
- Aspirate the blank and samples and record the concentration of Pb in sample solutions.

6. **Calculation :-**

Pb in ppm = (sample reading-blank reading) x Final dilution wt. of sample.

7. **Precautions:-**

- If mixing or grinding is necessary during sample preparation, porcelain mortar should be used if possible .
- Prepared sample should be stored in polyethylene bag or plastic container.
- Avoid sifting in preparation of samples to prevent metallic contamination or segregation of Pb.
- Carefully clean new glass, plastic and chemical ware with hot 10% sodium hydroxide solution followed by hot Nitric acid, and should be used for only Pb determinations.
- Only glass double-distilled or de-ionized water must be used for sample preparation and other purposes.
- Lead nitrate should be atleast 99% pure. It may be purified as follows:-

  Dissolve 20-50 gms in minimum of hot double distilled water and cool with the stirring. Filter crystals, redissolve and recrystallize. Dry crystals at 100-110ºC to constant weight. Cool in desiccator and store in tightly stoppered bottle.

Reference:-

AOAC (1990) 973. 35, 935.50)

Chapter 23

Standard Operating Procedures for Testing of Microbiological Quality of Milk

1. **General:-** As the microbiological quality of raw milk effects the quality and shelf life of the processed milk and milk products, it is important that the processor evaluates the quality of raw milk to ensure that only the milk of high quality is accepted in the plant for further processing. Standard Operating Procedures (SOPs) for various examinations of microbiological quality of milk are as follows:-

2. **Standard Plate Count (SPC) of Milk**

   Classically SPC procedures are used to determine the Total Plate Count (TPC) or Aerobic Plate Count (APC) or Total Viable Count (TVC). SPC is the standard method to which other screening tests are compared.

   2.1 Procedure:-

   2.1.1 Preparation of Diluent-Phosphate Buffer Solution Stock Solution:-

   Potassium dihydrogen phosphate 42.5g.

   Water (Distilled) 1000ml

   Dissolve the salt in 500ml of distilled water. Adjust the pH using IN NaoH or HCl solutions so that after sterilization it is 7.2 at 25ºC. Dilute to 1000ml. Distribute in screw capped sample bottles. Sterilize them at 121ºC for 15 minutes. Store the stock solution under refrigeration.

   2.1.2 Bench Solution:-

   Add 1ml of the stock solution to 1000ml of water for use as diluent. Dispense 9 ml of the diluent into test tubes. Stopper the tubes and sterilize by autoclaving at 121ºC for 15 minutes. The sterilized diluent could be stored at a temperature between 0ºC to 5ºC for one month in conditions which do not allow any change in its volume.

2.1.3 Preparation of Medium:-

Plate Count Agar or Standard Method Agar with the following composition is used:

Tryptone	5.0g
Yeast extracts	2.5g
Glucose	1.0g
Agar	15g
Distilled water	1000 ml.
pH	7.0 $\pm$ 0.2 at 25ºC

Prepare the medium by dissolving the ingredients in distilled water. Adjust the pH of the medium, so that it would be 7.0 $\pm$ 0.2 at 25ºC after sterilization for 15 minutes at 121ºC

If the commercially available media are used follow the manufacturers instructions for preparing the media.

2.1.4 Preparation of Test Samples:-

2.1.4.1 Raw Milk in Sample bottles:-

Remove from the closure any material that may contaminate the sample. The top of the unopened sample containers may be wiped off with a sterile cloth or paper towel saturated with 70% alcohol. Agitate the sample thoroughly so that micro-organisms are distributed as evenly as possible by rapidly inverting the container 25 times. Foaming should be avoided or foam allowed to disperse. The interval between mixing and removing the test portion shall not exceed 3 minutes.

2.1.4.2 Pasteurized milk in pouch:-

Clean the exterior of the pouch under tap water so as to remove any dirty material adhering to it. Shake the pouch to remove excess water and wipe off the traces of water adhering to the pouch with a tissue paper. Wipe the corner, where the packet is to be cut, with a tissue/cotton soaked in 70% alcohol to avoid external contamination from the pouch and allow the traces of alcohol to evaporate. Cut the sterilized corner of the pouch with a scissors (sterilized by dipping in alcohol followed by flaming to burn off the alcohol) and transfer about 100ml of milk to a clean sterile and dry sample bottle.

Agitate the sample thoroughly so that micro-organisms are distributed as evenly as possible by rapidly inverting the container 25 times. Foaming should be avoided or foam allowed to disperse. The interval between mixing and removing the test portion shall not exceed 3 minutes.

2.1.5 Preparation of decimal dilutions:-

Remove 1ml of the test sample with a pipette and add to 9ml diluent. Shake this primary dilution using a mechanical shaker for 5 to 10 seconds.

Transfer 1ml of the primary dilution into another tube containing 9ml of sterile diluent avoiding contact between pipette and the diluent. A fresh pipette should be used for each dilution.

Mix carefully, either by aspirating 10 times with a fresh pipette, or in a mechanical shaker for 5-10 seconds to obtain a 10^{-2} dilution.

If necessary repeat this operation using the 10^{-2} dilutions and further dilutions to obtain 10^{-3}, 10^{-4} etc. dilutions until the appropriate number of micro-organisms has been obtained.

Generally, 2nd and 3rd dilutions are sufficient for pasteurized milk, while one may require 5th, 6th or even 7th dilution for raw milk. However, the exact dilutions to be plated out for total counts are to be decided by one's own experience depending on the normal microbial load of the sample.

The time between the preparation of the initial suspension and the mixing of dilutions and the media shall not be more than 15 minutes.

2.1.6 Inoculation and Incubation:-

Take two sterile petridishes, using a sterile pipette, transfer to each dish 1ml of appropriate dilution of the test sample, say 10^{-5} for raw milk and 10^{-2} for pasteurized milk. Take two other sterile petridishes. Using a fresh sterile pipette, transfer to each dish 1ml of the subsequent decimal dilution of the test sample i.e. 10^{-6} and 10^{-3} dilutions for raw and pasteurized milk samples, respectively. If necessary, repeat the procedure with further dilutions of the test sample, using a fresh sterile pipette for each decimal dilution.

Pour about 15ml of the plate-count medium at $45^{\circ}C \pm 0.5^{\circ}C$ into each petridish. The time between the end of the preparation of the initial suspension and the mixing of dilutions and media shall not be more than 15 minutes carefully mix the inoculum with the medium and allow the mixture to solidify.

After complete solidification invert the prepared dish and incubate in the incubator set at $37^{\circ}C$ for $48h \pm 3h$. After the incubation count, using the colony counting equipment, the colonies in each dish contains not more than 300 colonies.

2.2 Calculation and expression of results:-

2.2.1 General case:- Dishes containing between 15 and 300 colonies. Retain the dishes containing not more than 300 colonies at two consequent dilutions. It is necessary that one of these dishes contain at least 15 colonies.

Calculate the number N of microorganisms present in the test sample as a weighted mean using the following equation:-

$N = \Sigma C\{(n1 + 0.1\ n2)\ d\}$

Where:

ΣC = is the sum of the colonies counted on all the dishes retained from two consecutive dilutions, and where at least one dish contains a minimum of 15 colonies.

n1 = the number of dishes retained in the first dilution.

n2 = the number of dishes retained in the second dilution

d = dilution factor corresponding to the first dilution. Round the result calculated to two significant figures. Express the result as a number between 1.0 and 9.9 multiplied by 10^x where x is the appropriate power of 10.

Examples- No. of colonies at the first dilution (10^{-2}): 168 and 215

No. of colonies at the second dilution (10^{-3}) : 14 and 25

$N = 168+215+14+25/\ \{2+(2 \times 0.1)\}\ 10^{-2}$

$= 422/0.022$

$= 19181$

= 19000 (rounding the result to two significant digits), which can be expressed as 1.9×10^4 per ml or g of the product.

2.2.2 Estimated Counts:-

2.2.2.1 Case where each dish retained contains less than 15 colonies. :-

If each of the dishes retained at two consequent dilutions contains less than 15 colonies, calculate the estimated NE of microorganisms using the above given equation.

Example:-

No. of colonies at the first dilution (10^{-4}); 3 and 5.

No. of colonies at the second dilution (10^{-5}); 0 and 1.

NE = $3+5+0+1/2.2 \times 10^{-4} = 9/2.2 \times 10^{-4} = 40000 = 4.0 \times 10^{4}$

2.2.2.2 Case of two dishes (test sample or first dilution retained)

Containing less than 15 colonies:-

If the two dishes corresponding to the test sample or from the first dilution inoculated or retained, contain less than 15 colonies, calculate the estimated number NE of micro-organisms present in the test sample as an arithmetical mean of the colonies counted on the two dishes using the following equation.

NE = $\Sigma C/V \times n \times d$

Where:

ΣC is the sum of the colonies counted on the two dishes.

V is the volume of inoculum applied to each dish, in milliliters.

n is the number of dishes retained (in this case n=2),

d is the dilution factor of the first dilution inoculated or retained (d=1) when the undiluted liquid product (test sample) is used.

Round off the result as given above.

Example:

No. of colonies at the first dilution (10^{-2}) retained: 12 and 13

NE = 12 + 13/1x2x10^{-2} = 25/0.02 = 1250

By rounding off the result the estimated number NE of micro-organisms is 13000 or 1.3x10^{3} per millilitre of product.

2.2.2.3 Case of two dishes (test sample or first dilution) containing no colonies:-

If the two dishes corresponding to the test sample or the first dilution inoculated or retained, do not contain any colonies express the result as follows:-

- Less than 1/d of micro-organisms per millilitre. Where d is the dilution factor of the first dilution inoculated or retained (d=1) when the undiluted liquid product (test sample) is used).

(i.e. if the two dishes from the test sample do not contain any colonies, report the result as less than 1 micro-organism per millilitre

Similarly, if the two dishes from the 10-1 dilution contains no colonies, report the result as less than 10 per millilitre.

2.2.2.4 Case of two dishes in the last inoculated dilution containing less than 300 colonies:-

Where only the two dishes containing the last inoculated dilution contain less than 300 colonies, calculate the numbers of micro-organisms present in the test. Sample as an arithmetical mean of the colonies counted on the two dishes, using the following equation.

$N = \Sigma C / v \times n \times d$

Where

ΣC is the sum of colonies counted on the two dishes, and where at least one contains a minimum of 15 colonies.

V is the volume of the inoculum applied to each dish, in milliliters.

n is the number of dishes retained (in this case n = 2).

d is the dilution factor corresponding to the dilution retained.

2.2.3 Express the result as follows:-

Number N' of micro-organisms per millilitre.

Example:

At the last dilution (10^{-4}) inoculated; 120 and 130 colonies.

N = 120 + 130/ 1x2x 10^{-4} = 250/ 0.0002 = 1250000

By rounding off the result the number 'N' of micro-organisms is 1300000 or 1.3×10^{6} per millilitre of product.

2.2.4 Interpretation:-

(a) Counts greater than 35000 per ml of pasteurized milk indicate unsatisfactory conditions. Presence of many pinpoint colonies on the plates indicates thermophilic contamination from the plant.

(b) The following standards are suggested as a guide for grading of raw milk on the basis of the plate counts if given in table -1

Table-1
Microbiological Standards for Grading of Raw Milk

Sr.No.	SPC / ml	Grade
1	Not exceeding 2,00,000	Very good
2	2,00,000 - 10,00000	Good
3	10,00000 - 50,00000	Fair
4	Over 50,00000	Poor

3. Coliform Count

The coliform group of bacteria comprises all aerobic and facultative anaerobic, gram-negative, non-spore forming rods able to ferment lactose with the production of acid and gas at 30ºC, 35ºC or 37ºC within 48 hrs. One source of these organisms is the intestinal tract of warm-blooded animals, certain bacteria of non-fecal origin are also members of this group. Typically, these bacteria are classified in the genera Escherichia, Enterobacter, and klebsiella. The presence of these coliforms in dairy products is suggestive of unsanitary conditions or practices during production processing or storage.

The coliform estimates are performed on raw milk to determine the degree of contamination during milk production. While the tests on pasteurized milk are useful to detect post-pasteurization contamination.

3.1 Procedure:

3.1.1 Preparation of Diluent-Phosphate Buffer Solution:-

Same as in SPC

3.1.2 Preparation of Medium

Violet Red Bile Agar (VRBA) medium with the following composition is used:-

Peptic digest of Animal Tissue	7.0g
Yeast Extract	3.0g
Bile Salts Mixture	1.5g.
Lactose	10.0g
Sodium Chloride	5.0g
Neutral Red	0.03g
Crystal Violet	0.002g
Agar	15.0g
Distilled Water	1000 ml.
Final pH	7.4 ± 0.2 at 25ºC

Prepare the medium by dissolving the ingredients in distilled water. Bring to boil. Allow it to boil for 2 minutes immediately cool the medium in the waterbath, set at 45ºC. Avoid overheating. Consequently do not autoclave the medium.

If the commercially available media are used follow the manufacturers instructions for preparation.

Use the medium within three hours of its preparation.

3.1.3 Preparation of Test Samples:

Same as in SPC

3.1.4 Inoculation and Incubation:- Take two sterile petridishes. Using a sterile pipette, transfer to each dish 1ml of the milk sample. Take two other sterile petridishes. Using a fresh sterile pipette, transfer to each dish 1ml of the subsequent decimal i.e. 10^{-1} dilution of the test sample.

If necessary, repeat the procedure with the further dilutions of the test samples, using a fresh sterile pipette for each decimal dilution.

In case, pasteurized milk, generally test-sample as such, 10^{-1} dilution are required to be plated out. Whereas for raw milk 2nd, 3rd and 4th dilutions may be tried out initially and then decide the appropriate dilutions required.

Pour about 15ml. of the VRBA medium at 45ºC $\pm$ 0.5ºC, into each, petridish. The time between the end of the preparation of the initial suspension and the mixing of dilutions and media shall not be more than 15 minutes. Carefully mix the inoculum with the medium and allow the mixture to solidify. After complete solidification, pour about 4 ml of the VRBA medium at 45ºC $\pm$ 0.5º onto the surface of the inoculated medium. Allow solidification of the medium, invert the prepared dishes and incubate them at 37ºC for 24h ± 2h. Retain the dishes containing not more than 150 colonies, whether characteristic or not for counting. Count the characteristic colonies in each of the retained dishes.

Characteristic colonies of coliforms are dark or purplish red colonies having a diameter of 0.5 mm or greater and sometimes surrounded by reddish zone of precipitated bile.

3.2 Calculation and Expression of Result:-

3.2.1 **General Case:** Dishes containing between 15 and 150 colonies:- Retain the dishes containing not more than 150 characteristic colonies at two consequent dilutions. It is necessary that one of these dishes contains at least 15 characteristic colonies.

Calculate the number (N) of coliforms per ml or per gram of product.

N = SC/{(n1 + 0.1 n2) d}

Where;

SC = is the sum of the characteristic colonies counted on all the dishes retained.

n1 = the number of dishes retained in the first dilution.

n2 = the number of dishes retained in the second dilution.

d= dilution factor corresponding to the first dilution.

Round the result calculated to two significant figures. Express the result as a number between 1.0 and 9.9 multiplied by 10^x where x is the appropriate power of 10.

Examples:

No. of characteristics colonies at the first dilution

(10^{-2}); 83 and 97.

No. of characteristics colonies at the second dilution (10^{-3}): 13 and 8

N= (83 + 97 + 13+8) / (2 + 0.1x2) 10^{-2}

= 9136

Rounding off the result gives 9100, which can be expressed as 9.1 x 10^3 per ml or g of the product.

3.2.2 Estimated Counts

3.2.2.1 Case where each dish retained contains less than 15 characteristic colonies:- If each of the dishes retained at two consequent dilutions. contains less than 15 characteristic colonies, calculate the estimated NE of coliforms using the above given equation.

Example:

No. of characteristic colonies at the first dilution (10^{-4}): 3 and 5.

No. of characteristic colonies at the second dilution (10^{-5}): 0 and 1

NE = 3+5+0+1/2.2x10^{-4} = 9/2.2 x 10^{-4} = 40,000 = 4.0x10^4

3.2.2.2 Case of two dishes (test sample or first dilution retained) containing less than 15 colonies.:-

If the two dishes corresponding to the test sample or from the first dilution inoculated or retained, contain less than 15 colonies, calculate the estimated number NE of microorganisms present in the test sample as an arithmetical mean of the colonies counted on the two dishes using the following equation.

NE = SC/ v x n x d

Where:

SC = is the sum of the colonies counted on the two dishes.

V= is the volume of inoculum applied to each dish in millilitres:

n= is the number of dishes retained (in this case n=2)

d= is the dilution factor of the first dilution inoculated or retained (d= 1 when the undiluted liquid product (test sample) is used).

Round off the result as given in above.

Example:

No. of colonies at the first dilution (10^{-2}) retained; 12 and 13.

$$NE = \frac{12 + 13}{1 \times 2 \times 10^{-2}} = \frac{25}{0.02} = 1250$$

By rounding off the result the estimated number NE of micro-organisms is 1300 or 1.3×10^3 per millilitre of product.

3.2.2.3 Case of two dishes (test sample or first dilution) containing no colonies:-

If the two dishes corresponding to the test sample or the first dilution inoculated or retained, do not contain any colonies express the result as follows:

- Less than 1/d of micro-organisms per millilitre where d is the dilution factor of the first dilution inoculated or retained (d=1 when the undiluted liquid product (test sample) is used).

If the two dishes from the test sample do not contain any colonies, report the result as less than 1 micro-organism per millilitre.

Similarly, if the two dishes from the 10^{-1} dilution contains no colonies, report the result as less than 10 per millilitre.

3.3 **Interpretation:-** Absence of coliform bacteria in 1/100 dilution (0.01 ml) in the case of raw milk and 1/10 dilution (0.1ml) in the case of pasteurized milk is accepted as criterion of satisfactory quality.

4. Direct Microscopic Somatic Cell Count (DMSCC)

4.1 **General:-** The direct microscopic method as regard somatic cell count is referred to as the direct microscopic somatic cell count (DMSCC). The number of somatic cells in raw milk provides a measure of the presence and the extent of mastitis or certain other abnormal milk secretions. DMSCC is applied as one of the officially recognized procedures for confirming the somatic cell counts, which were previously estimated by one of several screening tests. Test results are reported in actual counts of individual somatic cells.

4.2 **Procedure:-** There are five steps in the estimation of somatic cell count of milk by direct microscopic method. These are:

4.2.1 Estimation of Single Strip Factor (SSF)

4.2.2 Preparation of Milk Film.

4.2.3 Defatting, Fixing and Staining of the Smear.

4.2.4 Examinations and Counting of Stained Films.

4.2.5 Calculation and Expression of Results.

4.2.1 Estimation of Single Strip Factor (SSF):-

Place a clean stage micrometer on the stage of the microscope. Locate the micrometer scale under low power magnification (10x). Focus sharply and centre the scale in the microscopic field. Relocate the slide under oil immersion objective after placing a drop of immersion oil on to the 1mm scale of the stage micrometer. Focus sharply and center the scale in the centre of the microscopic field. Place the left edge of the scale on the left edge of the microscopic field. From left to right count the number of divisions which span the centre of the microscopic field.

Determine diameter (d) of the microscopic field in, mm

Calculate the SSF by using the following formula SSF= 10000/(area of the single strip) (in mm^2)

Where;

Area of the single strip = 11.28xd (mm^2) - if circle is used

= 10xd(mm^2) - if square is used

(d is the diameter of the microscopic field as determined above)

4.2.2 Preparation of Milk Film:-

Warm the milk sample to 40ºC immediately before transferring them to slides. Shake the sample thoroughly by rapidly inverting the container 25 times. Foaming should be avoided or foam allowed to disperse. The interval between mixing and removing the test portion shall not exceed 3 minutes.

Take out 0.01ml milk using a pipette or calibrated loop. Discharge the entire volume of milk onto the centre of the $1cm^2$ area (in the form of circle or square) of the slide. Spread the milk uniformly over the entire one square centimeter area. Dry films at 40 to 45ºC within five minutes using a hot plate or in an incubator.

4.2.3 Defatting, Fixing and Staining of the Smear:-

Dip the slides for ten minutes in a jar containing Newman's stain. Remove and drain off the excess stain and allow the slides to dry thoroughly. Rinse the slides in water until all the surplus dye is washed away. Drain and air-dry before examining the film under microscope.

4.2.4 Examination and Counting of stained Films:-

Examination:- Examine each film with an oil immersion objective after placing one drop of immersion oil on the film. Count only those somatic cells with an identifiable stained nucleus. For polymorphonucleated cells, count as a cell any that has two or more discernible nuclear lobes, for other somatic cells, count any that has a nucleus that appears to be essentially intact. If there is doubt about a cell, which may in fact be only a fragment, do not count.

4.2.4.1 Counting-field Wide Single Strip Method:-

This counting method uses as boundaries a single strip that runs the width of the microscopic field and across the diameter (if circle) or side (if square) of the milk film.

To make a single strip count, focus on the film edge in the oil immersion field that appears to be at the maximum horizontal or vertical excursion. Traverse the entire diameter of the milk film, counting all the cells within the strip and also the cells touching one edge of the strip. Do not count somatic cells that touch the other edge. During scanning of the strip, continually make fine focusing adjustments.

4.2.5 Calculation and Expression of Results:-

Compute somatic cells count as follows and express the result as direct microscopic somatic cell count (DMSCC) per ml of milk.

DMSCC per ml = No. of somatic calls in a single strip x SSF

Example:

Assuming an SSF of 4665; if an analyst counts 102 cells in a field wide strip, the count is computed as follows:

DMSCC per ml = 102 x 4665 = 475830

Round off the answer to two significant figures. The reported DMSCC becomes 480000 per ml.

4.3 **Interpretation:-** The presence of more than 500000 somatic cells/ml of milk is indicative of mastitis.

5. **Veterinary Drug Residues:-**

5.1 **General:-** Treatment of dairy cattle to prevent or cure diseases with drugs such as antibiotics, insecticides etc. will lead to the presence of residues of drugs in the milk for a certain period of time. The presence of these residues will have harmful effects if consumed by people and may lead to emergence of resistant strains of organisms. It may also seriously affect the manufacture of fermented dairy products like dahi, misti dahi, cheese etc. Therefore it is necessary to screen the milk supplies for the presence of drug residues.

The Delvotest is one of the tests used to detect the presence of drug residues in milk. The principle of the method involves germination and growth of spores of a specific bacteria (Bacillus-stearo thermophilus var. calidolactis) embedded in agar upon the addition of nutrients and milk. If milk is free of inhibitory substances, the growth of these spores produce acid, which changes the colour of the agar from purple to yellow. However, if milk contains certain inhibitory substances, these will diffuse into the agar medium and prevent the growth and subsequently the acid production by the bacteria and the medium remains purple in colour.

5.2 **Procedure:-** Follow the protocol of the manufacturer of the kit supplied along with each kit. Report the result as negative for antibiotics, if the colour of the medium changes to yellow and positive for the presence of antibiotics, if the colour remains purple after the specified incubation period.

Delvotest can give false positive reactions, if milk has added neutralizers, formalin and hydrogen peroxide.

Therefore, in case of positive result as evidenced by no change of purple colour, test the sample under reference for the presence of neutralizers, formalin and hydrogen peroxide.

The samples, which will be positive for Delvotest but negative for the presence of these additives, shall be sent to the designated laboratory for determining the presence and the levels of specific antibiotics viz benzyl penicillin, ceftiofur, sulfadimidine, neomycin, streptomycin, gentamycin and oxytetracycline residues. Take all the precautions for the dispatch of these samples as suggested for dispatch of samples for albendazole and others.

5.3 **Precautions:-** Store the kits upright in dark at a temperature of 6ºC to 15ºC. Milk samples shall be tested immediately after collection. If it is not possible store them refrigerated and complete the testing on the same day of sample collection.

Hands shall be washed and dried thoroughly before starting the test. Use a clean table.

5.4 **Additional Information:-** If found positive for the presence of inhibitory substances in the screening test and negative for the presence of added neutratizers, hydrogen peroxide and formalin proceed as follows for determining the presence of betalactam antibiotics and/or sulfonamides.

Add 0.1 ml of a solution of 10 IU of penicillinase per ml of water to 1.0ml of milk sample. Shake well and let it stand for 15 minutes at room temperature.

Add 0.1ml of an aqueous solution containing 1mg of sodium salt of para aminobenzoic acid to 100ml. of milk sample. Shake well and let it stand for 15 minutes at room temperature.

Add 0.1 ml of a solution of 10 IU of penicillinase per ml of water and 0.1ml of an aqueous solution containing 1 mg of sodium salt of para aminobenzoic acid to 1.0ml. of milk sample. Shake well and let it stand for 15 minutes at room temperature.

Repeat the Delvotest with the penicillinase, para-aminobenzoic acid and penicillinase and para aminobenzoic acid-treated milk sample as described above and report the result as follows:-

- Yellow colour in the penicillinase treated sample; confirmation of betalactam residues.
- Purple colour in the penicillinase treated sample-drug residues other than beta lactams.
- Yellow colour in the para amino benzoic acid-treated samples-confirmation of sulfonamides residues.
- Purple colour in the para aminobenzoic acid-treated samples for drug residues other than sulfonamides residues.
- Yellow colour in the para amino benzoic acid and penicilinase treated samples for confirmation of both penicillin and sulfonamides residues.
- Purple colour in the penicillinase and para aminobenzoic acid-treated samples for drug residues other than penicillins and sulfonamides residues.

5.5 **Interpretation:-** After adding 0.1 ml of milk sample, the ampoule is incubated for 2.5h at 63ºC to 66ºC. In the absence of anti-microbial substances, the whole of the solid medium turns yellow (negative), whereas it remains purple in the presence of sufficiently high concentrations of antibiotic. In case of doubtful results, the medium turns slightly yellow.

Chapter 24

Preparation of Media

1. **General :-** Various culture media (liquid, semi-solid and solid) are used in a bacteriological laboratory for the cultivation of micro-organisms, for studying their growth and physiological characteristics and for determining the number of viable organisms. In addition to natural media like milk several artificial media, containing different nutrients required for the growth of organisms, are employed for the purpose. The preparation of common culture media for use in bacteriological work is given as under.

2. **Adjustment of reaction (pH) of media.**

   2.1 **Introduction –** The reaction (pH) of the medium is one of the most important factors influencing the growth of micro-organisms. Most species of bacteria grow well when the reaction is neither acidic nor alkaline i.e. neutral (pH 7.0). Yeasts and moulds require a pronounced acid environment (pH 4 to 5) for their growth. In the preparation of culture media for the cultivation of micro-organisms adjustment of the reaction (pH) of the medium to the required level is a very important step. The pH of the medium is commonly estimated colourimetrically by adding a few drops of an indicator solution to the medium and matching the colour developed with a standard colour disc (corresponding to a definite pH) in a comparator. By adding a few drops of acid or alkali the pH of the medium is brought to the required level. In this title the estimation and adjustment of pH of phosphate buffer and peptone solutions is given.

   2.2 **Apparatus :-**

   2.2.1 pH Comparator.

   2.2.2 Standard pH colour discs with range from 6.0 to

7.6 pH. (Bromothymol blue indicator disc) B.D.H. or Hellige.

2.2.3 Test tubes (6" x 5/8").

2.2.4 Pipettes, 1ml and 10ml (graduated).

2.2.5 Bromothymol blue indicator solution (0.04% aqueous solution prepared by dissolving 0.1g of bromothymol blue powder in 16 to 18 ml of N/100 NaoH and adding distilled water to make up the volume to 250ml).

2.2.6 NaoH solutions (N and 0.1N strength).

2.2.7 Hydrochloric acid solutions (N and 0.1N strength).

2.2.8 Phosphate buffer solution (unknown pH).

2.2.9 Peptone water (0.5%) prepared by dissolving 0.5 grams of bacteriological peptone in 100ml of distilled water.

2.3 Procedure:-

2.3.1 Add 5ml of phosphate buffer solution to each of two clean tubes.

2.3.2 To one add 5 drops of standard B.T.B. indicator solution. Mix well by inverting the tube.

2.3.3 Using the comparator with the B.T.B. disc place the tube containing the indicator in one compartment and the other tube containing only the buffer solution in the other compartment against the colour disc. Turn the disc to give a reading of pH 6.2. If the colour of the test solution containing the indicator does not match with the colour standard in the disc and is more yellowish it means it is more acidic and the pH is below 6.2. Then take out the test tube, and add drop by drop 0.1N NaoH solution with a graduated 1ml pipette till the colour developed in the solution just matches with the standard colour disc corresponding to pH 6.2. If the colour of the solution is more greenish or shows a blue tinge it means that the solution is more alkaline. Then add, drop by drop 0.1 Hcl solution till the colour matches with that of standard disc (pH 6.2). Note the amount of NaoH or Hcl which has been added to the solution in the tube to bring the pH to the desired level.

Calculate the amount of 0.1N NaoH or 0.1Hcl required to be added to the rest of the solution. Add the required to be added to the rest of the solution. Add the required quantity of alkali or acid to the bulk solution. In order to avoid too much dilution, it is preferable to add calculated amounts of Normal NaoH and Hcl solutions corresponding to 0.1N NaoH or Hcl solution. Mix the solution well. Again find the pH of the solution as above. Repeat if necessary addition of acid or alkali until the desired pH of 6.2 is obtained.

2.3.4 Adjust the reaction of the buffer solution to pH 7.2 by using the same procedure as above.

2.3.5 Adjust the reaction of the given peptone solution also to pH 6.2 and 7.2 on the same lines as above.

2.3.6 Record and tabulate the results and show all the calculations for the adjustment of pH of phosphate buffer and peptone water.

3. Preparation of nutrient broth and nutrient agar .

3.1 Introduction – Nutrient broth is a liquid medium commonly used for the cultivation of aerobic organisms and also as a basal medium for a variety of physiological tests. Addition of agar to the broth gives a solid medium (nutrient agar) used for cultivation of several bacterial species and for determination of viable bacterial numbers.

3.2 Materials –

3.2.1 Beef extract (Lab Lemco)

3.2.2 Peptone

3.2.3 Agar (granular or shreaded)

3.2.4 Sodium chloride

3.2.5 Distilled water.

3.2.6 Sterilized test tubes, bottles and flasks.

3.3 Procedure –

3.3.1 Composition :-

Beef extract (Lab lemco)	3g
Peptone	5g
Sodium chloride	5g
Distilled water	1000 ml.
Agar powder or shred agar	15g.

(To be added only for nutrient agar)

(In the case of some commercial brands of agar 18 to 20g will be required to ensure rapid setting).

3.3.2 Weigh the ingredients and dissolve in 800ml of distilled water by steaming for 10 minutes.

3.3.3 Cool to 50ºc and adjust the pH to 7.2 according to method as given above. Steam again for 5 minutes. Filter through cotton pad.

3.3.4 Adjust the volume to 1000ml.

3.3.5 Dispense into sterilized test tubes, bottles or flasks.

4. Preparation of potato dextrose agar.

4.1 Introduction – Potato dextrose agar is a selective medium used for the cultivation of yeasts and molds. The use of a potato extract promotes growth of yeasts and molds, and the low pH (3.5) helps to inhibit the growth of bacteria while favouring the growth of yeasts and molds.

4.2 Materials :-

4.2.1 Raw potato

4.2.2 Dextrose

4.2.3 Agar

4.2.4 Sterile tartaric acid (10% solution)

4.2.5 Sterilized test tubes (6" x ¾").

4.3 Procedure –

4.3.1 Composition of medium –

Potato	200g
Dextrose	20g
Agar	20g
Water	1000ml.

(Final pH - 3.5).

4.3.2 Peel the potatoes and cut into small cubes.

4.3.3 Take 200g of potato cubes and 500ml of water in a flask and steam for one hour.

4.3.4 Mash by stirring well.

4.3.5 Strain through fine muslin and make up the filtrate to 1000ml.

4.3.6 Add 20g agar and steam again for 30 mts.

4.3.7 Filter, make up the volume, add 20g dextrose and tube in 10ml quantities.

4.3.8 Sterilize at 15 lb pressure for 20 mts.

4.3.9 Adjustment of pH to 3.5 should be made at the time of pouring plates.

4.3.10 Melt and cool to 45ºc two tubes of the medium. Adjust the reaction of the medium to pH 3.5 by adding a few drops of sterile 10% tartaric acid. The same amount of tartaric acid may be added to aseptically 10ml of medium in each tube at the time of pouring plates.

5. Preparation of Mac Conkey's bile salt broth and agar and desoxycholate agar.

5.1 Introduction :

This is a selective medium used for the cultivation of coliform organisms. The presence of sodium taurocholate (bile salt) in the medium lowers surface tension thereby favouring the growth of coliform bacteria. Hence it is of great value in the detection and isolation of coliform organisms. The medium is used for performing the presumptive coliform test for milk and water samples. Production of acid (indicated by the medium turning red or yellow according to indicator used) and gas in the Durham tube is taken as evidence of coliform contamination. Sodium Desoxycholate agar is also widely used for enumeration of coliform bacteria. This media is known for its selectiveness in favouring the growth of coliform organisms by producing characteristics red colonies against the agar background. This media has been reported to be more reliable, for estimating coliform content in raw and pasteurized milk than others.

5.2 Materials.

5.2.1 Sodium glyco or taurocholate or bile salt.

5.2.2 Peptone

5.2.3 Lactose

5.2.4 Sodium chloride

5.2.5 Distilled water

5.2.6 Andrade's indicator solution (0.5% aqueous solution of acid fuchsin).

5.2.7 Brom cresol purple solution (1% aqueous solution).

5.2.8 Durham's tubes (13/8" x 5/16").

5.2.9 Test tubes (6" x 5/8")

5.3 Procedure -

5.3.1 Preparation of Mac Conkey's Broth -

5.3.1.1 Composition

Sodium taurocholate	5g
Peptone	20g
Lactose	10g
Sodium chloride	5g
Distilled water	1000ml.

Andrade's indicator (or 2.5ml of brom cresol purple solution)

(Final pH - 7.4)

5.3.1.2 Weigh and dissolve the ingredients except lactose in 1000ml. of water.

5.3.1.3 Steam for 30 minutes. Adjust the reaction to pH 7.4.

5.3.1.4 Steam again for another 10 minutes. Add lactose

5.3.1.5 Filter through muslin cloth.

5.3.1.6 Add Andrade's indicator to give clear amber colour or brom cresol purple to give a light purple colour.

5.3.1.7 Dispense 8ml into sterilized 6" x 5/8" test tubes in which are placed inverted Durham tubes.

5.3.1.8 Sterilize at 15lb pressure for 20 minutes.

5.3.1.9 Incubate the tubes at 37ºC for 48 hours to detect and weed out any contaminated ones.

5.3.2 Preparation of Mac Conkey's agar -

This medium is used for the selective cultivation and enumeration of coliform bacteria. It is prepared by adding 15g of agar to the Mac Conkey broth medium, dispensing into test tubes and then sterilizing (autoclaving at 15lb for 20 minutes).

5.3.3 Desoxycholate agar (Difco)

5.3.3.1 Composition :-

Bacto peptone	10g

Bacto lactose	10g
Sodium desoxycholate	1g
Ferric citrate	1g
Sodium citrate	1g
Dipotassium hydrogen phosphate	2g
Neutral red	0.03g
Agar	15g
Distilled water	1000ml
pH	7.2

5.3.3.2 Dissolve the ingredients and allow to stand for 5 minutes and mix thoroughly to a uniform suspension.

5.3.3.3 Heat slowly with stirring and boil it for five minutes.

5.3.3.4 Filter, boil for 10 minutes and distribute in test tubes.

5.3.3.5 Store in the refrigerator if not immediately used.

(Inhibitory to Gram +ve cocci and spore forming organisms but not to Gram –ve enteric bacilli).

Chapter 25

Bacteriology of Air

1. **General :-** Air is not a natural environment for the growth and reproduction of micro-organisms but a few types of organisms (e.g. aerobic spore formers, micrococci, mold spores) may, however, be found in association with dust particles suspended in the air. In the vicinity of cattle sheds, dairies and fermentation industries the atmosphere may carry several types of bacteria, yeasts, mold spores, and bacteriophages. Air is also known to be a medium for the transmission of infectious micro-organisms (e.g. viruses). The air inside a bacteriological laboratory is frequently required to be sterilized by irradiation or fumigation by bactericidal chemicals (aerosols). Bacterial content of fresh air outside the town may vary from 0.25 to 6.0 per c.ft. A crowded office may show 20 to 65 per c.ft. In a service hospital 200-1000 per c.ft.

2. **Estimation of bacteria in air :-**

    2.1 **Introduction :-** A knowledge of the numbers and types of micro-organisms present in the atmosphere inside dairies, cattle byres and food factories, is very important for controlling contamination of milk, milk products and other materials. It is required to estimate the bacterial content of air in a milking shed, butter storage rooms, milk product packaging rooms, and in bacteriological laboratory.

    2.2 **Materials:-**

    2.2.1 Sterile petridishes

    2.2.2 Nutrient agar tubes

    2.2.3 Sabouraud agar tubes

3. **Procedure :-**

3.1 Pour two plates with melted nutrient agar and two plates with sabouraud agar. Allow the media to set and harden.

3.2 Remove the tops from the plates.

3.3 Place one plate of each medium on the floor of the milking shed etc. and allow them to be exposed for 5 minutes. Immediately replace the tops on the plates.

3.4 Place one plate of each medium on the laboratory bench, allow the plates to be exposed for 15 minutes and immediately replace to tops.

3.5 Incubate the nutrient agar plates at 37ºc for 2 days and the sabouraud agar plates at 22ºc for 3 to 4 days.

3.6 At the end of the incubation period count the number of colonies in each plate.

3.7 Prepare smears of typical bacterial colonies in each plate and examine for morphology by Gram's Stain.

3.8 Examine mold-like colonies using the low power objective of a microscope.

3.9 Record your observations and interpret.

4. **Interpretation** – Number of organisms per c.ft. per minute should not exceed one for satisfactory atmosphere.

Chapter 26

Assessing Sterility of Plant and Equipment

1. **Introduction** – The sanitary condition of plant and equipment including storage tanks, vats, coolers, pipelines, heat exchangers, agitators fittings etc. is tested by applying the swab technique. In this method a sterile cotton or wire-gauge swab (wetted with buffer solution) is rubbed over the surface of the equipment and the bacterial cells removed from the surface are transferred into sterile buffer solution and their numbers estimated by standard plate count.

2. **Materials** –

   2.1 Apparatus and materials for taking standard plate counts as milk.

   2.2 Test tubes containing 25ml Ringer's or Phosphate buffer solution. (If the equipment to be tested is sanitized by treatment with chlorine it is necessary to incorporate 0.05 per cent of sodium thiosulphate in the solution. When a quaternary ammonium compound is used for the purpose add 0.4% lecithin and 1.0% Tween 80 or Tween 20 to the solution).

   2.3 Swab consisting of cotton wool or wire gauge.

   2.4 Washed and sanitized tank or vat.

   2.5 Preparation of Swab – The swab consists of wire-gauge or cotton wool wound round the end of a metal wire (preferably of stainless steel). Take a piece of metal wire (35 cm. long x 2.6mm diameter), formed into a loop at one end leaving a straight length of 30cm and notched at the other end to hold the gauge or cotton wool. Wind a piece of 50mm unmedicated ribbon gauge 150mm long) or non-absorbent cotton wool round the notched end of the metal wire over a length of 50mm and secure the swab with thread. Place the swab in 25ml of the Ringer's or phosphate buffer solution in a test tube, plug with cotton wool and sterilize by autoclaving at 120ºc for 15 minutes.

3. **Procedure :-**

3.1 Press the swab with a rolling motion against the side of the glass tube to remove the excess liquid and take it out of the tube.

3.2 Rub the swab with heavy pressure back and forth over the area to be examined so that all parts of the surface are treated twice. The swab should be rotated so that all parts of it make contact with the surface. The swabbing should be repeated over 9 spots, of 10x5 cm area in different parts of the surface so that the total area covered comes to 900 sq.cm wherever possible. To facilitate swabbing over required areas a thin metallic mask (12.5 x 7.5 cm) with a cut-out area of 10 x 5 cm in the centre, may be sterilized and used for guidance.

3.3 After rubbing the required area return the swab to the solution in the tube in which it was originally placed.

3.4 Allow the swab to be immersed in the liquid for 5 minutes and mix by swirling the swab vigorously in the solution 6 times. Remove the swab after expressing the excess liquid by pressing against the side of the test tube.

3.5 Mix the solution (swab sample) thoroughly by rotating the tube between the palms of the hands.

3.6 Prepare 1/10 dilution of the sample using a 9 ml dilution blank.

3.7 Transfer 1ml portions of the swab sample as well as the 1/10 dilution into duplicate plates and add agar medium.

3.8 Incubate the plates at 37ºC plus or minus 0.5ºC for 48 hours.

3.9 At the end of the incubation period count the plates having 30 to 300 colonies and find the colony count per ml of the swab sample. This number multiplied by 25 gives the colony count of the total area swabbed from which the count per 900 sq.cm area can be calculated.

3.10 Express the result as colony count per 900 sq.cm area of the surface of equipment.

4. **Interpretation** - The following standards are suggested as a guide for judging the sanitary conditions of the equipment.

Colony Count per 900 sq.cm area	Sterility
Less than 5,000	Satisfactory
5,000 to 25,000	Fairly satisfactory
More than 25,000	Unsatisfactory

Chapter 27

Assessing Sterility of Milk Bottles

1. **Introduction** – In market milk plants, where milk is pasteurized, chilled and filled in bottles for distribution, it is of utmost importance to prevent the recontamination of milk from improperly cleansed and sanitized milk bottles. Each batch of bottles should be examined soon after washing for satisfactory sanitization. Four or more bottles, selected at random from each batch, should be tested for sterility by the rinse technique. In this method the bottle or other container is rinsed with a known amount of sterile Ringer's or phosphate buffer solution to remove the bacterial cells remaining on the surface and their number estimated by taking standard plate counts of the rinse solutions.

2. **Materials :-**

   2.1 Apparatus and materials required for standard plate count of milk.

   2.2 Sterile Ringer's solution or phosphate buffer (20ml in test tubes or vials) (If the solution is to be used for testing bottles rinsed with chlorinated water, crystalline sodium thiosulphate should be added to the Ringer's solution before sterilization to give a concentration of 0.05 per cent).

   2.3 Samples of washed bottles closed with sterile rubber bungs.

3. **Procedure :-**

   3.1 Add 20ml of Ringer's or buffer solution to the bottle and replace the bung. The same amount of solution can be used for rinsing bottles of different sizes.

   3.2 Hold the bottle horizontally in the hands, rotate gently 12 times in one direction and also shake lengthwise 12 times so as to wet the whole of the internal surface.

3.3 Allow the bottle to stand for 15 to 30 minutes and again gently rotate 12 times to wet the internal surface thoroughly.

3.4 Transfer 1ml and 5ml portions of the rinse solution to two sets of petridishes in duplicate, pour the agar medium, allow the medium to settle and incubate the plates at 37ºc plus or minus 0.5ºc for 48 hours.

3.5 At the end of 48 hours count the number of colonies selecting those plates showing colonies between 30 and 300.

3.6 The average of the counts in duplicate plates multiplied by 20 (in the case of plates containing 1ml of the rinse) and by 4 (if the plates contained 5ml of the rinse) gives the colony count per bottle.

3.7 Report the results in terms of colony counts per bottle.

4. **Interpretation** – Colony count of more than one colony per ml of the capacity of the bottle is an indication of unsatisfactory sterility. The result in terms of colony counts per bottle is given in table-1.

Table-1

Capacity	Colony counts per Bottle	Sterility
250ml	250 and below	Satisfactory
250ml	More than 250	Unsatisfactory
500ml	500 and below	Satisfactory
500ml	More than 500	Unsatisfactory

Chapter 28

Assessing Sterility of Milk Cans

1. **Introduction** – Milk cans, pails and other large containers may add heavy contamination to milk if they are not cleaned and sanitized properly. The general condition of the utensils e.g. presence of dents open seams, milk stones and moisture, is also important as these defects lead to pockets of contamination which may escape destruction during sanitization treatments. The sanitary condition (sterility) of milk can is tested by the rinse technique. The can should be examined immediately after washing and sanitizing treatment.

2. **Materials :-**

   2.1 Apparatus and materials required for standard plate count of milk.

   2.2 Flasks containing 500ml of sterile Ringer's or buffer solution.

   2.3 Milk cans (washed and sanitized)

3. **Procedure :-**

   3.1 Make a visual inspection of each can and note the presence of any dent, open seams, poor lids, rusty spots on the interior surface, moisture, dirt and film and scale of milk solids.

   3.2 Pour 500ml of sterile Ringer's or buffer solution over the inside of the lid into the can. If the can is sterilized by chlorine treatment it is necessary to incorporate sodium thiosulphate into the Ringer's or buffer solution to give a concentration of 0.05 per cent.

   3.3 Replace the lid, lay the can on its side and roll it to and fro so that it makes 12 complete revolutions.

3.4 Pour the rinse sample from the can into a sterile bottle or flask.

3.5 The rinse sample should be examined, immediately. Otherwise it should be placed in a refrigerator (2ºC or 3ºC) and examined not later than 24 hours.

3.6 Invert the container slowly three times to mix the rinse sample.

3.7 Transfer 1ml of the sample into duplicate plate.

3.8 Prepare 1/10 dilution of the sample using a 9ml dilution blank and transfer 1ml of the dilution into another set of plates.

3.9 Pour agar medium into the plates and after the medium is set incubate the plates at 37ºC plus or minus 0.5ºC for 48 hours.

3.10 After the incubation period, remove the plates and count the colonies, plates having colonies from 30 to 300 should be used for counting.

3.11 The average of the counts in duplicate plates (multiplied by 10 in the case of 1/10 dilution) represents the colony count per ml of the rinse sample.

3.12 Express the results in terms of colony count per ml (colony count per ml of rinse x 500).

4. **Interpretation** - Calculate the colony count per litre capacity of the can (colony count per can divided by its capacity in litres) and grade the sanitary condition of the can as follows:-

Colony Count per litre capacity	Sterility
Less than 1,000	Satisfactory
1,000 to 5,000	Fairly satisfactory
More than 5,000	Unsatisfactory

Note: Any defect in the can noted during visual inspection also indicates general unsatisfactory conditions of the can.

Chapter 29

Microbiological Analysis of Fermented Milk Products

1. **General :-** The quality of dahi and other fermented milk preparation is likely to be seriously impaired if undesirable micro organisms (e.g. coliform bacteria, yeasts, molds) gain entry during manufacture from milk, starter-culture, unsterile equipment and atmosphere. The products are particularly susceptible to contamination by yeasts and molds which find the acidic environment favourable for their growth. The presence of such organisms in appreciable numbers is therefore, an indication or a reflection of insanitary conditions of manufacture.

2. **Examination of dahi -**

   2.1 Introduction - Dahi is a typical product of lactic fermentation of milk and is very popular in India. The procedure for examining this product for presence of undesirable contaminants is described in this title.

   2.2 Presumptive Coliform Test

   2.2.1 Materials

   2.2.1.1 Sample of dahi

   2.2.1.2 Dilution blanks (9 or 99ml).

   2.2.1.3 Pipettes (1, 1.1, and 10ml).

   2.2.1.4 Mac Conkey's broth tubes with Durhams fermentation tubes.

   2.2.1.5 Mac Conkey's agar

   2.2.1.6 Petridishes for plating.

2.2.2 Procedure :-

2.2.2.1 Mix the sample thoroughly by shaking vigorously so that uniform consistency is obtained.

2.2.2.2 Prepare dilutions of 1:10 and 1:100 using sterile 1ml pipettes and 9ml blanks.

2.2.2.3 Transfer 1g portions of dahi and its dilutions (1/10 and 1/100) into Mac Conkey's broth tubes in triplicate.

2.2.2.4 Incubate the tubes for 24 hours at 37ºC and observe for the production of acid and gas. The production of acid is indicated by change of colour of medium from purple to yellow in the case of brom cresol purple and orange to pink in the case of Andrade's indicator. Production of gas is observed in the Durham's tubes which may be partially or completely filled with gas.

2.2.2.5 If no change is observed, incubate for another period of 24 hours and record the observation.

2.2.2.6 Tabulate the results in the following manner –

Date	Culture Dilution	Reaction of the Mac Conkey's Broth tubes	Positive Coliform +ve	Remark
		A G A G A G		
	1 2 3 4			

A - Acid G - Gas

2.2.3 Interpretation :-

Dahi of good quality should give a negative coliform test in 1/10 dilution. Positive test in

1/10 or higher dilutions indicates insanitary conditions or manufacture.

2.3 Yeast and mold count :-

2.3.1 Introduction :-

Dahi prepared under hygienic condition from freshly heated milk is usually free from yeasts and molds. Hence presence of large number of yeasts and molds in dahi indicates improper methods of heating of milk or defective starter culture used during preparation of curd or poor sanitary conditions of manufacture, packaging and storage.

2.3.2 Materials

2.3.2.1 Sample of curd or dahi

2.3.2.2 Dilution blanks

2.3.2.3 Potato dextrose agar

2.3.2.4 Pipettes (1.0 and 10.0 ml)

2.3.2.5 Petridishes

2.3.3 Procedure :-

2.3.3.1 Mix the sample of dahi, thoroughly by agitation and if necessary use a sterile glass rod for complete mixing.

2.3.3.2 Weigh 10 gram of dahi in sterilized petridish and transfer the dahi to 90 ml dilution blank warmed to 45ºc and mix the contents by shaking. This gives 1:10 dilution.

2.3.3.3 For preparing and plating 1:10 and 1;100 dilutions follow the same procedure as described for milk.

2.3.3.4 Transfer 1ml of 1:10 dilution to duplicate petridishes for plating in 1:10 dilution.

2.3.3.5 Adjust the reaction of potato dextrose agar to pH 3.5 by adding calculated amount of sterile tartaric acid solution at the time of pouring plates.

2.3.3.6 Pour the melted and cooled agar and mix the contents well. Allow the agar to cool and set.

2.3.3.7 Invert and incubate the plates at 21ºC or 25ºC for 5 days. If molds grow fast and develop into large colonies, plates may be incubated for 3 days only.

2.3.3.8 Count the number of colonies in the plates and compute the number of colonies per gram of dahi.

2.3.3.9 Report the results as yeast mold plate count per gram of dahi.

2.3.4 Interpretation :- If yeast and mold count exceeds 100 per gram of dahi poor quality is indicated.

Chapter 30

Starter Cultures

1. **Introduction:-** Selected strains of lactic acid bacteria (eg. Str. lactis, Str. cremoris, Str. thermophilus, Leuc. dextrainicum, L. bulgaricus, L. acidophilus, L. helveticum,) are used, singly or in combination of two or more species as starter cultures in the manufacture of several milk products like dahi, yoghurt, acidophilus milk, cultured milk, ripened cream butter and cheese. Their function is to produce acid and flavour compounds and bring about the coagulation of milk and other desired changes in milk or cream. They also play an important role in the ripening of cheese. Since these cultures are liable to become inactive or get contaminated with bacteriophages and, other undesirable organisms extreme care is necessary in their maintenance and propagation in the dairy laboratory. The following experiments are designed to make the students familiar with the routine maintenance and propagation of starter culture and checking their purity and activity for use in manufacture of products.

2. **Starter Activity tests:-**

   2.1 **Introduction:-** During the continued propagation of starter cultures, the cultures are likely to become weak and lose their activity and viability due to contamination by undesirable micro-organisms, such cultures will not be useful for manufacture of dairy products of good quality. It is therefore, necessary to test periodically the activity of the cultures and their ability to produce acid or flavour compounds. Since, the activity of the culture is judged by observing the rate of acid production by the organisms, their ability reduce methylene blue or resazurin added to milk and also form volatile flavour compounds (diacetyl) in milk. In order to insure the activity of starter, three tests i.e. acidity test, dye reduction test and creatine test are tested and described as under.

2.1.1 Acidity test:

2.1.1.1 Procedure;

(i) Mix the culture thoroughly and transfer 0.3ml to 10ml. of skim milk.

(ii) Incubate the tube at 37ºC for 3½ hours.

(iii) Transfer the entire contents, using 5ml. distilled water in to a 100ml. conical flask and add a few drops of phenolphthalein.

(iv) Titrate against 0.1 N NaoH solution

(v) Calculate the acidity in terms of per cent lactic acid by multiplying vol. of 0.1 NaoH used in titration with 0.09 multiplying factor.

2.1.1.2 Interpretation:- Acidity of 0.35% or more indicates that the culture is satisfactory.

2.1.2 Dye Reduction Test:

2.1.2.1 Procedure:

(i) Mix the culture thoroughly and transfer 1ml to 10ml of sterile skim milk. Mix the contents well.

(ii) After that 10ml content is transferred to the other sterile test tube add one ml MBR-solution and incubate at 37ºC in waterbath

(iii) Note the time of decolourisation.

2.1.2.2 Interpretation:- The quality of the culture is judged on the basis of the time taken for reduction of methylene blue as given in table-1

Table-1

Methylene blue reduction of the culture

S.No.	Reduction Time of methylene blue	Grade
1	Less than 35 minutes	Very good
2	Between 35 and 50 minutes	Satisfactory
3	More than 1 hour	Poor

2.1.3 Creatine Test (test for diacetyl)

2.1.3.1 Material (i) Dahi and butter cultures (ii) Test tubes (iii) 30% potassium hydroxide solution (iv) Creatine

2.1.3.2 Procedure:- (i) Mix the culture well and transfer about 2 to 2.5ml into a test tube. Add a small quantity of creatine (1-2 mg).

(ii) Add equal quantity (2 to 2.5ml) of the alkali solution.

(iii) Mix thoroughly by shaking the tube and allow to stand for 10 minutes.

(iv) Observe the formation of a pink band on the surface of milk and the time taken.

2.1.3.3 Interpretation:- Formation of a pink band at the surface in about 10 minutes due to production of diacetyl in the culture indicates that the culture is active in producing flavour.

Chapter 31

Bacteriological Analysis of Cream

1. **Introduction:** Milk cream being susceptible to contamination from various sources during its production, if it is exposed to warm temperature during storage, rapid growth of micro-organisms (bacteria, yeasts and moulds) takes place resulting in the development of souring and other defects. The same types and numbers of bacteria commonly found in milk are also found in cream. Hence the methods of bacteriological analysis of cream and the interpretation of results are similar to those adopted for milk.

2. **Standard Plate Count of Cream**

   2.1 **Introduction:** The standard plate count gives an estimate of the number of viable organisms present in cream. Because of higher total solids content, viscocity and additional equipment used in the manufacture of cream, slightly larger numbers of organisms are allowed in the grading of cream by the agar plate method.

   2.2 **Materials :-**

   2.2.1 Sample case: metal or wooden insulated box with space for placing sample bottles and crushed ice keeps the samples at 32º-40ºC.

   2.2.2 Plungers.

   2.2.3 Sampling dippers or measures.

   2.2.4 Sample bottles (4 oz glass stoppered, wide mouthed dust proof bottles).

   2.2.5 Apparatus required for standard plate count as milk.

   2.3 **Procedure :-**

   2.3.1 **Sampling of Cream** – Because of its perishable

nature special asceptic techniques are required for the collection and handling of cream samples for bacteriological analysis.

2.3.1.1 When the cream is not in frozen condition, agitate the cream in a can or vat with the help of a sterile plunger to mix the sample thoroughly with the help of a sampling dipper or measure transfer about 100 ml of cream into a sample bottle, label promptly and transfer it to the sampling case.

2.3.1.2 If the cream is in a frozen state, temper the entire can or vat to 40ºC with the lid on. After the entire mass has attained the temperature, agitate the sample as in the previous case and transfer 100 ml to the bottle. Label and transfer to the case.

2.3.1.3 In case of sour cream samples also, follow the same procedure as indicated above for bacteriological analysis.

2.3.2 Plating the Samples :-

2.3.2.1 Because frozen cream or cream held at low temperatures cannot be pipetted easily. Temper the samples by holding the bottle in a waterbath at 37º-40ºC for sufficient time (about 15 minutes) to bring the samples into a melted homogeneous state. Mix the samples by shaking each container.

2.3.2.2 Using sterile pipettes transfer 1ml of well mixed sample to the first dilution blank (9ml or 99ml).

2.3.2.3 Prepare from this further dilutions as needed.

2.3.2.4 Transfer 1ml each of the final and next lower dilution to duplicate petridishes. Mark the dilution on the plates.

2.3.2.5 Pour 12-15ml of melted, cooled agar and mix the contents thoroughly and allow to cool and set.

2.3.2.6 When cooled, invert the plates and incubate at 37ºC for 48 hours.

2.3.2.7 Count the colonies with the aid of colony counter and tally counter after selecting the set of plates containing colonies between 30 and 300.

2.3.2.8 After counting the colonies, calculate the standard plate count per ml of sample.

2.4 Interpretation:- The following standards are recommended for judging the quality of raw and pasteurized cream, as given in table-1

Table-1

Raw and pasteurized cream standards of standard plate count

Standard plate count per ml	Quality
(a) Raw Cream	
Less than 4,00,000	Good
4,00,000 to 20,00,000	Fair
More than 20,00,000	Poor
(b) Pasteurized Cream	
Less than 60,000	Good
60,000 to 2,00,000	Fair
More than 2,00,000	Poor

3. Presumptive Coliform test for Cream :-

3.1 Introduction - Determination of coliform contamination in raw cream, as in the case of raw milk, has limited application, since small numbers of coliform bacteria may always enter into cream through various sources during normal handling and can multiply quickly into larger numbers under favourable conditions. However, excessive numbers of coliforms in raw cream are undesirable as they generally indicate in sanitary conditions of production and handling. Coliforms do not survive pasteurization temperature and their presence in pasteurized cream generally indicates re-contamination. Hence the presumptive coliform test is particularly useful in checking post-contamination in pasteurized cream. The test may be carried out by using either solid or liquid media.

3.2 Materials :-

3.2.1 Samples of pasteurized cream (obtain the samples aseptically).

3.2.2 Mac Conkey's agar.

3.2.3 Dilution blank (9 or 99 ml).

3.2.4 Pipettes 1.0 and 1.1 ml.

3.2.5 Petridishes.

3.3 Procedure :-

3.3.1 Prepare the sample as described in 2.3.2.1.

3.3.2 Mix the samples well and transfer 1ml to 9ml dilution milk. This gives a dilution of 1 : 10.

3.3.3 Prepare further dilution if necessary in another 9ml of dilution water. This gives a dilution of 1:100.

3.3.4 Transfer 1ml portions of each dilution into duplicate petridishes previously labelled.

3.3.5 Pour 10-15 ml of melted, cooled Mac Conkey's agar into the plates and mix the contents by gentle rotation.

3.3.6 Allow the agar to cool and solidify.

3.3.7 Pour the second layer of the same agar (5-10ml) over the first layer and spread evenly by gentle rotation. Allow to cool.

3.3.8 Invert the plates and incubate them at 37ºC for 24/ 48 hours.

3.3.9 At the end of 24 hours remove the plates for counting colonies. Plates showing no colonies should be incubated for a further period of 24 hours and then examined.

3.3.10 Examine the plates for the presence of typical coliform colonies (raddish in the case of Mac Conkey's agar of at least 0.5mm diameter). The presence of such colonies constitutes a positive test.

3.3.11 Also count the number of such colonies in duplicate plates and report the average of the counts as number of coliform bacteria per ml of cream.

3.4 Interpretation :- The following tentative standards are suggested for judging the quality of cream

Sample	Coliform test	Grade
Raw Cream	Absent in 1:100 dilution	Satisfactory
Pasteurized Cream	Absent in 1:10 dilution	Satisfactory

4. Methylene blue reduction test for cream :-

4.1 Introduction :- Methylene blue reduction test could be conveniently used to judge the bacteriological quality of cream. As in the case of milk, grading of cream based on this test has been suggested. Because of the higher solids content and viscosity of the product the use of triple concentration of the dye has been recommended. The test could also be carried out after appropriate dilution of the cream. Methylene blue reduction times of cream obtained by this method correspond fairly well with the plate counts or direct microscopic counts of the samples.

4.2 Materials :-

4.2.1 Samples of cream.

4.2.2 Methylene blue solution of triple concentration (The solution is prepared in a similar way as in the case of milk testing, except that the concentration of the dye is increased three times).

4.2.3 Test tubes 6" x 5/ 8".

4.2.4 Rubber Corks (No. 2 size) to fit into the above tubes.

4.2.5 Pipettes (1.0 and 10.0 ml)

4.2.6 Beakers, forceps and flasks.

4.3 Procedure -

4.3.1 If frozen or cooled to low temperature, warm the sample by placing in a waterbath at 40ºC for a few minutes. After the sample has melted, mix well by gentle agitation.

4.3.2 Transfer 10ml of well mixed cream to a sterile test tube by means of a pipette.

4.3.3 Add 1ml of methylene blue solution and close the tube with a sterile rubber cork. Mix the cream and dye by inverting the tube a few times until a uniform blue colour is seen.

4.3.4 Place the tubes in the waterbath and record the time.

4.3.5 Observe the tubes after half an hour for any reduction. If no reduction has taken place replace the tubes in the waterbath.

4.3.6 Subsequently, observe after every one hour until the dye is reduced. Invert such of the tubes which have not shown any reduction and return them back to the waterbath for further observation.

Note: An interval timer may be used to remind the observer when successive examinations are to be made.

4.3.7 Note the time taken for complete decolourisation and express the data in terms of complete half hour intervals.

4.4 Interpretation - The standards of mythylene blue reduction time are suggested for judging the quality of cream as given in table-2

Table-2

Standards of mythylene blue reduction time

Methylene blue Reduction Time in Hours	Quality of Cream
5 and above	Very good
3 and 4	Good
1 and 2	Fair
½ and below	Poor

Chapter 32

Ice–Cream Analysis

1. **Determination of weight per litre and over-run in Ice cream**

1.1 General – The determination of over-run in frozen and hardened ice-cream is a somewhat complex problem to solve due to the fact that the weight of mix is unknown and has to be determined before calculations can be made.

The over-run in ice-cream depends upon the amount of air whipped into the mix during the freezing process. In this test, the volume of water and alcohol used corresponds with the volume of air originally contained in the ice-cream and the difference between the sum of these two and the capacity of the flask is equivalent to the volume occupied by the sample.

1.2 Apparatus :-

1.2.1	Analytical Balance	- For weighing accurately to 0.001 g.
1.2.2	Beaker	- 400 ml.
1.2.3	Volumetric flask	- 250 ml.
1.2.4	Glass funnel	

1.3 Reagents :-

1.3.1 n – Amyl Alcohol – Sp. gr. 0.817.

1.4 Procedure :-

1.4.1 Weigh a unit of ice-cream and from it calculate the weight of the ice-cream per litre. For example, 200ml of a full carton of ice-cream can be obtained, the ice-cream carefully removed and the empty dry carton weighted. The difference in weights between the carton when filled and when empty is, therefore, the weight of 200ml of frozen ice-cream five times, this weight would then equal the weight of a litre. To determine the weight of the mix, as given below:-

1.4.2 Weigh and record the exact weight of a clean, dry 400 ml beaker. Into the beaker weigh exactly 130 g of the frozen ice-cream. Place the beaker in waterbath warmed to 49ºC and melt. Weigh and record the exact weight of a 250ml volumetric flask. Using a glass funnel, transfer 130g of melted ice-cream into the 250ml flask. Add exactly 10g of n-amyl alcohol to the flask and mix to break the surface tension of the melted ice-cream and release the incorporated air. Ten grams of n-amyl alcohol occupy a volume of 12.24 ml. Cool the flask with contents to 15.5ºc using a cold water of ice waterbath. Rinse the beaker containing melted mix with several small rinsings of distilled water, adding each rinse to the 250 ml flask. Again cool the flask with contents to 15.5ºC and using the final rinse water, bring the volume to 250 ml mark. The bottom of the meniscus should correspond with the mark when temperature is exactly 15.5ºC. Dry the outside of the flask and reweigh.

1.4.3 Calculate the weight in grams of the contents. Calculate the volume in millilitres occupied by the sample of ice-cream. Determine the specific gravity of the mix by dividing its weight (130g) by the volume in millilitres which it occupied. Determine the weight in grams per litre of mix by multiplying by the specific gravity.

Note: After weighing the unit of ice-cream it is essential to remove the carton carefully without tearing and dry it thoroughly before reweighing. If the ice-cream has been heat-shocked, accurate results cannot be obtained.

2. Determination of total solids:-

2.1 Preparation of samples for chemical Analysis

2.1.1 Plain Products :- Let the sample soften at room temperature. Because melted fat tends to separate and rises to the surface, it is not advisable to soften the sample by heating on waterbath or overflame. Mix thoroughly by stirring with spoon or egg beaker or by pouring back and forth between beakers.

2.1.2 Fruit nut and chocolate ice-cream containing insoluble particles :-

2.1.2.1 Use a mixer capable of comminuting product to fine, uniform pulp. Use 100-200g of sample to fill the cup of mixer full to about one-third. Melt the product at room temperature or in an incubator at 37ºC in closed container. Transfer entire contents to the mixture cup and mix until insoluble particles are finely divided (about 3-5 minutes for fruit ices and upto 7 minutes for nut ices). Alternatively, the product may be ground in a porcelain or glass pestle and morter.

2.1.2.2 Transfer the mixed sample to a suitable container for convenience in weighing. After weighing operations, return the remainder of the sample to the refrigerator, preferably at a temperature not exceeding -15ºC.

2.2 Apparatus :-

2.2.1 Flat Bottomed Dishes of nickel or other suitable metal not affected by boiling water, 7 to 8 cm in diameter and not more 2.5 cm deep, provided with short glass stirring rods having a widening flat end.

2.2.2 Sand

Which passes through 500 micron IS sieve and is

retained on 180 micron IS sieve. It shall be prepared by digestion with concentrated hydrochloric acid, followed by thorough washing with water. It shall then be dried and ignited to dull red heat.

2.2.3 Well-Ventilated Oven

Capable of operating at 102ºC air temperature.

2.3 Procedure :-

2.3.1 Heat the necessary number of metal dishes, each dish containing about 20g of prepared sand and a stirring rod, in the oven for about one hour. Allow to cool in an efficient desiccator for 30 to 40 minutes. Weigh accurately about 3g of the prepared sample of ice-cream into a dish. Saturate the sand by careful addition of a few drops of distilled water, and thoroughly mix the wet sand with ice-cream by stirring with glass rod, smoothing out lumps and spreading the mixture over the bottom of the dish.

2.3.1.1 Place the disk on a boiling waterbath for 20 to 30 minutes, then wipe the bottom of the dish and transfer it, with the glass rod, to the well-ventilated oven at 102 ± 1ºC. The bulb of the oven control thermometer shall be immediately above the shelf carrying the dish. Dishes shall not be placed near the walls of the oven, and should be insulated from the shelf by suitable silica or glass supports.

2.3.1.2 After 4 hours, remove the dish to an efficient desiccator, allow to cool as before and weigh. Replace the dish in the oven for a further period of 1 hour at 102 ± 1ºC, remove to the desiccator, and cool and weigh again. Repeat the process of heating, cooling and weighing for one hour till consecutive weighings agree to within 0.5 mg.

2.4 Calculation :-

2.4.1 From the loss in weight observed, calculate the per cent by weight of total solids. for ice-cream.

3-A Determination of fat (Rose Gottlieb Method)

3A.1 Apparatus :-

3A.1.1. Fat - Extraction Apparatus - Either of the following apparatus may be used.

(a) A fat-extraction tube conforming to the dimensions and capacities, 5 or 6, fitted with either a wash bottle top or a siphon, carrying the two tapes in a two-holed bark-cork and provided also with a ground-glass stopper or with a solid bark cork. The solid bark cork used to close the tube shall be sound, free from pores and channels which would allow leakage of solvent, and previously extracted with ether. The narrow tube with the hook-shaped lower end is a sliding fit in the cork and of such length that the opening at its lower end may be placed, if necessary, at a distance of 25mm from the bottom of the tube.

Note: Other modifications of the Rohrig tube of the same capacity may also be used.

(b) A Mojonnier fat extraction tube of closed with a solid bark or ground glass stopper.

3A.1.2. A well-ventilated electrically heated oven - set to operate at 98 to 100ºC.

3A.2. Reagents :-

3A.2.1. Concentrated Ammonia Solution - approximately 35 per cent m/m (sp. gr. 0.88)

3A.2.2. Ethyl Alcohol - 95 to 96 per cent (v/v).

3A.2.3. Diethyl Ether - Sp. gr. 0.720, peroxide free.

Note: Diethyl ether may be maintained free from peroxide by adding wet zinc foil (approximately 80 cm²/ litre, cut in strips long enough to reach at least half way up the container) that has been completely immersed in dilute acidified copper sulphate for one minute and subsequently washed with water.

3A.2.4. Light Petroleum - boiling range 40 to 60ºc.

3A.2.5. Mixed Solvent - Prepared by mixing equal volumes of ether and light petroleum.

3A.3. Procedure :-

3A.3.1 Weigh accurately 4.5g of the prepared sample to the fat extraction tube. Wash the sides of the tube with 2ml of hot water and mix by gentle swirling. Add 2ml of concentrated ammonia solution and mix thoroughly but without splashing the contents to the upper part of the tube. Heat in waterbath for 20 minutes at 60ºC with occasional shaking. Add 10ml of ethyl alcohol, mix well. Complete extraction of the fat is dependent on satisfactory mixing at each stage.

Add 25ml of ether, close the tube with the cork (or stopper) which is wetted with water before insertion, and shake vigorously for one minute. Remove the cork and, with 25ml of light petroleum, wash the cork and neck of the tube so that the washings run into the tube. Replace the cork again wetted with water, and shake vigorously for 30 seconds. (It is essential that the cork (or stopper) be wetted with water before each insertion and washed with solvent during each removal. Also, before each removal, to avoid spurting of the solvent, a slightly reduced pressure should be induced in the tube by cooling. Rubber stoppers shall not be used).

Allow the tube to stand until the ethereal layer is clear and completely separated from the aqueous layer, usually for not less than 30 minutes.

(Separation of layers may be achieved by the use of a centrifuge). Remove the cork and insert the siphon (or wash bottle) fitting so adjusted for length that the inlet is 2 to 3mm above the interface between the ethereal and aquesous layers, and transfer the ethereal layer to a suitable flash. Add 5ml of mixed solvent to the extraction tube, using it to wash the siphon or wash bottle fitting which is raised sufficiently to permit this but not removed, and the inside of the tube. Lower the fittings and transfer the solvent without shaking to the flask. Repeat this operation with a further 5ml of mixed solvent. Wash the tip of the siphon fitting into the flask with mixed solvent.

Remove the siphon fitting, and repeat the extraction of the milk residue, using 15ml of ether and 15ml of light petroleum, and repeat the subsequent operations as before. Use the ether to wash the inner limb of the siphon (or wash - bottle) fitting during its removal from the tube. Finally, repeat the extraction once more with 15ml each of ether and petroleum.

Distil carefully the solvents from the flask and dry the residual fat in the oven at 98 to 100ºC for one hour taking precautions to remove all traces of volatile solvent and cooling the flask to room temperature in a desiccator charged with an efficient desiccant. Repeat this procedure for periods of half an hour until successive weighings do not show a loss in weight by more than 1mg.

Extract completely the fat from the flask by repeated washing with light petroleum, allowing any sediment to settle before each decantation, dry the flask in the oven, cool and weigh as before. The difference in weights before and after the petroleum extractions, subject, to a correction if necessary, for the blank described below, is the weight of fat contained in the weight of milk taken.

Make a blank determination using the specified quantities of reagents throughout, and water in place of the ice-cream, and deduct the value found, if any, from the apparent weight of fat. A flask, similar to that used to contain the fat, shall receive the same heating and cooling treatments and shall be used as a counterweight.

3.B Determination of Fat by Gerber Method

3B.1 Apparatus :-

3B.1.1. Gerber centrifuge.

3B.1.2 Ice-cream butyrometer.

3B.1.3 Waterbath.

3B.1.4. Automatic measure of 10ml and 1ml.

3B.1.5. Weighing balance.

3B.2 Reagents -

3B.2.1. Dilute Gerber sulphuric acid (70:13 volume for vanilla ice cream and 94:6 volumes for chocolate ice-cream)

3B.2.2. ISO - amyl alcohol.

3B.3. Procedure :- Weigh accurately 5g of melted sample into the ice-cream butyrometer. Add 6ml of hot water (for dilution) and washing the sides of butyrometer. Take 10ml of diluted Gerber H_2SO_4 (70:13 volumes for vanilla ice-cream and 94:6 volumes for chocolate ice-cream) into the butyrometer. Add 1ml of iso-amyl alcohol. Insert the stopper, shake and centrifuge for 5 minutes at 1100 rpm. Keep the butyrometer at 65ºC in a waterbath for 5 minutes as is done for milk. Observe the fat column in the butyrometer stem and read the percentage.

4. Determination of Acidity :- The acidity of the ice-cream shall be determined before the addition of colouring matter.

4.1 Apparatus :- The apparatus shall be same as in milk.

4.2 Reagents :- The reagents shall be same as in milk.

4.3 Procedure – The procedure shall be the same as in milk but using 20g of the prepared sample being diluted with about 50ml of recently boiled cooled water.

4.4 Calculation

4.4.1 Titratable acidity (as lactic acid)

$$\text{Percent by weight} = \frac{9VN}{W}$$

Where;

V = Volume in ml of the standard sodium hydroxide solution required for titration.

N = Normality of the standard sodium hydroxide solution used, and

W = Weight in g of the product taken for the test.

5. Determination of Sucrose :–

5.1 Reagents :-

5.1.1 Fehling's Solution : Prepare by mixing equal volumes of solution A (copper sulphate solution) and solution B (potassium sodium tartrate solution) immediately before using.

20ml = approximately 40ml of 0.25 per cent invert sugar solution.

5.1.1.1 Standardization of Fehling's solution :

Pipette accurately 20ml of Fehling's solution prepared as above into a 250ml conical flask. Add from a burette a volume about 1ml less than the expected volume of standard dilute invert sugar solution which will reduce Fehling's solution completely (about 40ml) and sufficient water to bring the volume to 75ml at the commencement of boiling. Heat rapidly to boiling on asbestos gauge. Reduce the heat sufficiently to maintain slow but steady boiling, and in two minutes from the onset of boiling add 1ml of methylene blue solution. Add small quantities of the standard invert sugar solution until the indicator is decolourized. The titration must be completed in about three minutes, excluding air by maintaining ebullition, and refraining from rotating or shaking the flask throughout. After this preliminary titration a further titration or titrations should be carried out, adding practically the whole of the invert sugar solution required before commencing the heating and continuing the titration as before.

5.1.1.2 The colour change is best judged in good north daylight or its equivalent and it is recommended that the boiling should be carried out on a clean white asbestos gauge or a thin white silica tile.

5.1.1.3 From the volume of invert sugar used, calculate the equivalent of 20ml of Fehling solution in terms of milligram (x) of invert sugar.

5.1.2 Copper Sulphate Solution (Solution A)

Dissolve 34:639g of copper sulphate ($CuSO4-5H_2O$) in water, dilute to 500ml and filter, if necessary.

5.1.3 Potassium sodium tartrate (Rochelle Salt) solution (Solution B)

Dissolve 173g of potassium sodium tartrate and 50g of sodium hydroxide in water, dilute to 500ml and filter, if necessary, through asbestos.

5.1.4 Hydrochloric acid - density 1.18 at 20ºC approximately 12N.

5.1.5 Hydrochloric acid - 6.34N.

5.1.6 Alumina cream -

To a cold saturated solution of aluminium potassium sulphate in water add sufficient aqueous ammonia (density 0.880) stirring constantly to make the mixing alkaline to litmus paper. Wash with water, by decantation, allowing the gelatinous precipitate to settle thoroughly between each washing, until only a trace of sulphate remains in the washings.

5.1.7 **Neutral Lead Acetate Solution** - Prepare a concentrated solution of lead acetate in cold water, neutralize, if necessary, to litmus paper by adding acetic acid or sodium hydroxide, dilute to a density of approximately 1.25 at 20ºC and filter. This requires the solution of about 41g of lead acetate diluted to a final volume of 100ml, after neutralization.

5.1.8 Sodium oxalate solution - Saturated solution in water.

5.1.9 Methylene blue indicator - 0.2 per cent in water.

5.1.10 Sodium hydroxide - approximately N.

5.1.11 Standard invert sugar solution.

5.1.12 Stock invert sugar Solution -

Transfer 23.75g of sucrose dried overnight in a vacuum desiccator, to a one-litre graduated flask with the aid of about 100ml of water. Add 10ml of concentrated hydrochloric acid, mix thoroughly and allow to stand for at least three days at a temperature approximating 20ºC. Dilute to 1 litre. This stock solution contains 2.5g of invert sugar per 100ml and may be kept for up to four weeks without changing its concentration.

5.1.13 Standard dilute invert sugar solution -

5.1.13.1 Transfer 50ml of the stock solution, make just neutral to litmus paper with IN sodium hydroxide solution and dilute to 500ml immediately before use.

1ml = 2.5 mg Invert sugar.

5.1.13.2 The above procedure is recommended, but where it is necessary to use a rapidly prepared solution, transfer 1.1875g of dried sucrose to a 500ml flask with 75ml of water. Add 10ml of 6.34N of hydrochloric acid slowly while rotating the flask. Partially immerse the flask in a waterbath adjusted to 70ºC and when the temperature of the contents reaches 67ºC (which should take 2½ to 2 ¾ minutes), allow five minutes further heating by which time the temperature should have reached about 69.5ºC. Remove the flask and cool immediately by immersing in a water bath at 20ºC. When the contents of the flask have nearly reached 20ºC, make just neutral to litmus with IN sodium hydroxide solution and dilute to 500ml at 20ºC.

1ml = 2.5 mg invert sugar.

5.2 Procedure :-

5.2.1 Determination of Original reducing sugars :-

Place about 10g of the prepared sample of ice-cream accurately weighed (it is convenient to weigh by difference) into a 250ml conical flask and dilute with 150ml of water. Mix thoroughly the contents of the flask. Add neutral lead acetate drop by drop, mixing by rotating the flask until no further precipitate is formed.

Add one drop of alumina cream, again mix and allow to stand for a few minutes. Rotate the flask at intervals to ensure complete precipitation of protein. Add just sufficient sodium oxalate solution to precipitate any excess lead. Filter through a fluted 18cm No.1 filter paper into a 250ml graduated flask. Wash the precipitate and the paper thoroughly, with hot water collecting the washings in the flask. Cool the flask and contents and make up to the mark (solution C). Carry out titrations against Fehling's solution prepared and standardized as above.

5.2.1.1 If the weight of sample taken is w/g and the volume of sugar solution used in the titration is 5 ml, then the percentage of original reducing sugar in the ice-cream;

$$= \frac{25\ x}{S.W}$$

5.2.1.2 The concentration of reducing sugar in the filtered solution may be such that more than 55ml are required to reduce 20ml mixed Fehling's solution. In this case employ the following modification.

5.2.1.3 Pipette 10ml each of Fehling's solution prepared and standardized as above into the flask, add an accurately known volume of the standard dilute invert sugar solution and complete the titration with the prepared sugar filtrate as described under standardization of Fehling's solution. Subtract from the number of milligram of invert sugar found, the amount contained in the known volume of standard invert sugar solution added.

5.2.2 Determination of reducing sugars after Inversion:-

Transfer 50ml of the filtered solution (solution c) to a 200ml graduated flask, add 25ml of water and 10ml of 6.34N hydrochloric acid. Invert as described in the preparation of standard dilute invert sugar solution. Determine the total invert sugar as described in the determination of reducing sugars.

5.3 Calculation :-

5.3.1 Sucrose = [(reducing sugars after inversion) - (Original reducing - sugars)]

X 0.95.

6. Estimation of Total Colony Count (Standard plate count)

6.1 Preparation of Sample :-

6.1.1 To avoid any difficulties in obtaining representative test portions only melted samples may be used. For the purpose of melting, the frozen sample may be kept at room temperature or, if required, in a water bath at a temperature not exceeding 45°c for not more than 15 minutes. Thoroughly mix the samples before removal of test portion.

6.1.2 Measurement of Test Portion -

For greater accuracy, in view of variable over-runs and differences in density of mixes, it is preferable to use only gravimetric measurement in preference to the volumetric method. Using sterile pipette, aseptically transfer 11.0g (using a balance sensitive to 30 mg) of test portion directly in to dilution bottles containing 99ml of buffered distilled water or ringers solution.

6.1.3 Procedure - The procedure shall be the same as in the milk.

6.1.4 In such cases where constantly low standard plate counts are obtained, it may be desirable to occasionally incubate the plates at 5 to 7°c for 7 days instead of 37°c to determine the presence of psychrophillic organisms.

7. Determination of Coliform Counts -

7.1 General - Presence of coliform organism in frozen dairy products usually signifies improper processing, subsequent contamination from equipment, flies, personnel etc, or addition ingredients to mixes after pasteurization and generally insanitary conditions of handling. The coliform numbers can be measured by either of the two procedures employing solid media or liquid media. Many workers, however, prefer to use solid media because of higher reproducibility. The solid media also permit prompt confirmation of any doubtful colonies. In case of frozen products where fermentable carbohydrates other than lactose are present, positive coliform results must be confirmed to obviate false positive results due to the action of non-lactose fermenting non-coliform organisms.

7.2 Apparatus and Materials :-

7.2.1 The apparatus and materials shall be the same as in the milk. In addition, the following will be required.

(a) Balance, sensitive to 30 mg, with suitable weights.

(b) Sterile stainless steel spatula, suitable length.

7.3 Preparation of the sample and Measurement of Test Portion :-

7.3.1 The methods of preparation of the sample and measurement of test portion shall be the same as prescribed under standard plate count.

7.4 Procedure :-

7.4.1 Coliform Test with Solid Media - Transfer appropriate volume (0.1 to 1.0ml) of suitable dilution in to sterile plates. Add 10–15ml of violet-red bile agar or desoxycholate agar. With a view to increasing the sensitivily of tests, larger test portions (up to 4.0ml) may be used with 15-20ml of the medium. Quite often 10ml of 1:10 dilution is distributed in 2-4 plates using 15-20ml of the medium per plate. Mix the contents thoroughly and allow to solidify (within 5-10 minutes), add 3-4ml of the same melted medium in each plate as an over larger to completely cover the surface to inhibit formation of surface colonies, Invert and incubate for 18-24 hours at 37 + 0.5ºC.

7.4.1.1. Dark-red colonies measuring 0.5mm or more on uncrowded plates are considered to be coliform bacteria. Count such colonies and report as coliform colonies per g using two significant figures and substituting decimals for whole number where necessary.

7.4.1.2 If no coliform appears on plate(s) report as coliform colonies less than – per g inserting the number that would be reported. If only one coliform colony had been found from the total quantity plated, for example if 2g had been plated report as coliform colonies less than 0.5/g

7.5 Completed Test With Solid Media:-

7.5.1 In case of doubt, for example, in case of crowded plates, where coliform colonies may not have typical appearance as also to confirm presence of lactose fermenting coliform bacteria, promptly transfer typical colonies to lactose - broth or to brilliant green bile-broth tubes. Production of acid and gas within 48 hours confirms the presence of coliform organisms.

7.5.1.1 In such cases where typical, clearly isolated colonies are not present, streak growth on surface of violet-red bile agar or desoxycholate agar for isolation and transfer typical colonies to lactose broth or to brilliant green bile broth as above. Production of acid and gas within 48 hours confirms the presence of coliform organisms.

7.6 Interpretation :-

7.6.1 In properly operated plants, coliform counts of not more than 100g may be expected.

8. Phosphatase Test :-

8.1 Apparatus - The apparatus shall be the same as in the milk.

8.2 Reagents - The reagents shall be the same as in the milk.

8.3 Procedure :-

8.3.1 Use mix sample before nuts, fruits or flavouring and colouring materials have been added. While testing finished ice-cream, stain out any nuts or fruits before testing. Run an extra control with each individual sample to evaluate colour causes by flavouring materials. Proceed as in the milk and interprete the result as in the milk.

Note:- Another condition responsible for false positive tests in frozen dairy foods is related to whether the flavouring, primarily vanilla type, is added to mixes before or after pasteurization. When vanilla is added to mixes consisting in whole or in part of unpasteurized products and the mixes are then commercially pasteurized before testing, blue colour develops under conditions, of the phosphatase test. However, when vanillin is added to milk products which previously have been heated to 80ºC and the mixture is commercially pasteurized, the products are negative to phosphatase test. This difference is attributed to activity of residual enzymes on vanillin and may vary with the time interval allowed for enzyme activity, amount and nature of substrate, and case of liberating phenol from it. Obviously the intensity of blue colour is greater when vanillin is added before pasteurization instead of after, and may increase as the time interval is extended between its addition and the time of testing after pasteurization of milk or mix. By the same token, use of value obtained or a properly pasteurized vanillin free milk or mix to which vanillin is added after pasteurization and before applying the test may cause erroneously high phosphatase value which if used for comparision with the result on a vanillin free under-pasteurized product, would give rise to a false negative interpretation that the later product had been properly pasteurized. Since the presence of vanillin in mixes not previously heated to over 80ºC ordinarily may lead to false positive interpretations, it is essential that either a laboratory pasteurized sample of the mix be run as a control or that a control test substituting buffered water for buffered substrate be used on all samples giving a positive test. Blue colour developed by this control would be attributable to interference. If this colour is just equal in intensity to that of sample tube, the sample is judged pasteurized. If colour in sample tube is distinctly greater than that of control, under-pasteurization is indicated.

Chapter 33

Determination of Fat in Milk Products by the Gerber Method

1. **Milk Products:-**

1.1 Apparatus:- The apparatus given in 1.1.1 to 1.1.8 conforming to the provisions of IS: 1223 (Part 1)-1970-IS:1223 (Part-II)-1972 and IS: 1223 (Part-III)-1977 are required.

Note:- (i) Specification for apparatus for determination of milk fat by Gerber method; Part 1 bytyrometers and stoppers (first revision)

(ii) Specification for apparatus for determination of milk fat by Gerber method; Part-II Pipettes and Automatic measures (first revision).

1.1.1 Butyrometer:- 70 per cent scale - for estimating fat in cream.

1.1.2 Butyrometer, 6 per cent, 8 per cent and 10 per cent scale for estimating fat in milk powders.

1.1.3 Cheese butyrometer:- 40 per cent scale - for estimating fat in cheese.

1.1.4 10ml pipette or automatic measure - for sulphuric acid.

1.1.5 1ml pipette or automatic measure - for amyl alcohol.

1.1.6 Stoppers for butyrometers.

1.1.7 Centrifuge

1.1.8 Waterbath

1.2 Other Apparatus:-

1.2.1 Weighing balance suitable for weighing to 0.01g.

1.2.2 Small scoop-of suitable material (for example, aluminium, nickel or plastics). It is convenient if this scoop has a counterpoise.

1.2.3 Stemless funnel

1.2.4 Glass rod of such diameter (6mm is convenient) that the rod will pass through the hole at the base of the funnel.

1.2.5 Small camel hair brush

1.2.6 Wash bottles containing cold (30-40ºC) and hot (about 70ºC) distilled water (IS: 1070-1977).

1.2.7 Stoppered funnel.

1.2.8 Grater or pestle and mortar.

1.3 Reagents:-

1.3.1 Sulphuric acid shall have a density of 1.807 to 1.812g/ml at 27ºC corresponding with a concentration of sulphuric acid from 90 to 91 per cent by mass.

(a) The sulphuric acid shall be colourless or not darker than pale amber in colour.

(b) When diluted with distilled water to a density of 1.4 g/ml, not more than a very slight turbidity shall occur.

1.3.2 Amyl alcohol shall conform to Grade 1 of IS: 360-1964

1.4 Method for Cream :-

1.4.1 Mixing of sample: Stir the sample thoroughly but not so vigorously as to cause undue froth or churning. If the cream is very thick, warm to a temperature between 30ºC and 40ºC to facilitate mixing. Mix immediately before weighing the required amount of cream for the test.

Note:- If it is not possible to achieve complete mixing, the sample should not be tested by this method.

1.4.2 Method-1:-

(a) Weighing of sample: weigh 5.00 + 0.01g of sample into a dried and tared beaker of about 50ml capacity.

(b) **Addition of acid and alcohol:** Add in small quantities freshly prepared mixture of one volume of sulphuric acid and one volume of distilled water to the cream in the beaker. Dissolve the cream and transfer the mixture carefully, with the help of a dry funnel, to the butyrometer for testing cream. Rinse the beaker about six times with small quantities of the dilute acid to make sure that all cream has been transferred to the butyrometer. Altogether 18 to 20ml of the diluted acid should suffice to fill the butyrometer, leaving sufficient space for the addition of amyl alcohol and mixing. Add 1ml of amyl alcohol by means of the 1ml pipette or the automatic measure. Do not wet the neck of the butyrometer with alcohol.

(c) **Insertion of Stopper:** Close the neck of the butyrometer firmly with the stopper, without disturbing the contents. When a double ended stopper is used, screw it in until the widest part is at least level with the top of the neck. When a lock stopper is used, insert it until the rim is in contact with the neck of the butyrometer.

(d) **Mixing of Contents:** Shake the butyrometer carefully without inverting it until the contents are thoroughly mixed, the curd is dissolved and no white particles are seen in the liquid. Then invert the butyrometer a few times to mix the contents thoroughly.

Note:- When a large number of samples are to be mixed, shake the butyrometers in a protected stand until the contents are thoroughly mixed and no white particles are seen. Invert once or twice during the process.

(e) **Temperature adjustment:** Transfer the butyrometer quickly, with the bulb uppermost, into a waterbath having a temperature of 65 ± 2ºC and leave it there for not less than 3 minutes and not more than 10 minutes. Take care to have the water level in the bath above the top of the fat column in the butyrometer. Meanwhile, adjust the stopper so that the fat column should be on the scale after centrifuging.

(f) **Centrifuging:** Take the butyrometer out of the water, dry it with a cloth and transfer it to the centrifuge, placing two butyrometers diametrically opposite so as to balance the rotating disc centrifuge at the maximum speed for 5 minutes. Bring the centrifuge to stop gradually. Transfer the butyrometers, stoppers downwards, into a water bath having a temperature of 65 ± 2ºC and allow the butyrometers to stand in the waterbath for at least 3 minutes and not more than 10 minutes.

(g) **Reading of butyrometer:** Before taking a reading, adjust the position of the fat column to bring the lower end of the column onto a main graduation mark. When double ended stoppers are used, do this by slightly withdrawing the stopper and not by forcing it further into the neck. Note the scale readings, corresponding to the lowest point of the fat meniscus and the surface of separation of the fat and acid, the difference between the two readings gives the percentage by mass of fat in the cream. When readings are being taken, hold the butyrometer with the graduated portion vertical, keep the point read in level with the eye and then read the butyrometer to the nearest 0.5 per cent, that is, half the smallest scale division.

(h) **Second Centrifuging:** If after centrifuging there is not a sharp dividing line between the fat and the acid, or if the acid layer is not clear, repeat the temperature adjustment and centrifuging before taking the reading.

(i) **Precautions:** If a fluffy layer is observed at the base of the fat column in the butyrometer, reject the test. Examine the stopper to see if it is in good condition, repeat the test and take care to ensure that the curd is completely dissolved.

(i) If the fat column is so dark as to make reading difficult, reject the test and check the strength of the sulphuric acid.

(ii) If large number of samples have to be tested, it is preferable to use automatic measures for measuring the sulphuric acid and amyl alcohol, especially the latter, otherwise there is a possibility of injurious effects to the health arising from the inhalation of amyl alcohol vapours by the use of 1ml pipette.

(j) **Check reading:** Replace the butyrometer in the bath for another 3 minutes and then take a check reading of the butyrometer as rapidly as possible.

1.4.3 Method 2 :-

(a) **Addition of acid to butyrometer:** Transfer 10ml of sulphuric acid into the butyrometer by means of the 10ml pipette for sulphuric acid or the automatic measure, taking care not to wet the neck of the butyrometer with the sulphuric acid.

(b) **Weighing of sample** – Mix and immediately weigh 5 + 0.01g of sample into the butyrometer without soiling the neck, using any suitable form of support for the butyrometer on the balance.

(c) **Addition of Water and amyl alcohol:** Add about 6ml of the hot water (70ºC) to the butyrometer. Measure 1ml of amyl alcohol into the butyrometer by means of the alcohol pipette or automatic measure. Adjust the level of the contents to about 5mm below the shoulder by further additions of hot water. In no circumstances shall the amyl alcohol be added to the butyrometer before the cream.

(d) Follow the procedures described in 1.4.2 (c) to (j).

1.5 Method for Milk Powders:-

1.5.1 **Weighing of sample -** Mix the sample thoroughly and weigh 1.69 + 0.01g. of milk powder into the counterpoised scoop.

1.5.2 **Addition of acid to butyrometer:** Follow the procedure prescribed in 1.4.3 (a) using the butyrometer.

1.5.3 **First addition of water:** Add gently from the wash bottle sufficient cold water to form a layer about 6mm deep on top of the acid, allowing the water to flow down the side of the bulb.

1.5.4 **Addition of sample:** Insert the narrow end of the stemless funnel into the neck of the butyrometer. Transfer the contents of the scoop to the funnel, removing the last particles with the camel hair brush. Tap the funnel gently until most of the powder is in the butyrometer. Transfer the remaining powder in the funnel to the butyrometer with the aid of the glass rod and the camel hair brush. Remove the funnel.

1.5.5 **Addition of amyl alcohol:** Measure 1ml of amyl alcohol into the butyrometer by means of the 1ml pipette for amyl alcohol or the automatic measure. Do not wet the neck of the butyrometer with alcohol.

1.5.6 **Second addition of water:** Add hot water (70ºC) from the wash bottle until the butyrometer is filled to about 5mm below the shoulder, allowing all air entrained in the powder to escape.

1.5.7 Follow the procedures prescribed in [1.4.2 (c) to (j)]. Read the butyrometer to the nearest 0.05 per cent.

1.5.8 **Third Centrifuging:** Repeat the procedure described in 1.4.1 (h) if the reading obtained after the second centrifuging is higher than that obtained after the first centrifuging.

1.5.9 **Calculation of percentage of fat:** Multiply the scale reading of the butyrometer by the factor 20/3 to obtain the percentage of fat in milk powders.

1.6 **Method for Cheese :-**

1.6.1 **Preparation of sample:** Grate samples of hard cheese. Grind samples of soft cheese. Mix thoroughly.

1.6.2 **Weighing of sample:** Counter balance the stoppered funnel with its stopper inserted. Weigh 3 ± 0.01g of the sample into the funnel.

1.6.3 **Addition of acid to butyrometer:** Follow the procedure prescribed in 1.4.3 (a) using the cheese butyrometer.

1.6.4 **First addition of water:** Add gently from the wash bottle sufficient warm water (30 to 40ºC) to form a layer about 6 mm deep on top of the acid, allowing the water to flow down the side of the bulb.

1.6.5 **Addition of sample:** Insert the neck of the funnel containing 3g of cheese into the neck of the butyrometer. Withdraw the stopper from the neck of the funnel and transfer all the cheese to the butyrometer with the aid of the glass rod or spatula.

1.6.6 **Addition of amyl alcohol:** Follow the procedure prescribed in 1.5.5.

1.6.7 **Second addition of water:** Add warm water (30 to 40ºC) from the wash bottle until the butyrometer is filled to about 5mm below the shoulder.

1.6.8 Follow the procedures described in 1.4.2 (c) to (j). Read the butyrometer to the nearest 0.3 per cent, that is, one-third of the smallest scale division.

Note: For cheese containing more than 40 per cent of fat, 1.5g of cheese should be taken for the test and the butyrometer reading multiplied by 2.

1.7 Method for Curd:-

1.7.1 **Preparation of sample:** Mix thoroughly the curd for analysis of fat.

1.7.2 **Weighing the sample:** Weigh 100g thoroughly mixed curd into a clean, dry, 250ml of beaker.

1.7.3 **Addition of liquid Ammonia:** Add 5–10ml of (NH40H) liquid ammonia into a beaker and mix to a homogeneous mass.

1.7.4 **Addition of acid to butyrometer:** Follow the procedure prescribed in 1.4.3 (a) using the milk butyrometer.

1.7.5 **Addition of sample –** Measure 10.75ml of sample into the required butyrometer by means of the 10.75ml pipette, the temperature of sample, should be brought to approximately 27ºC when it is measured.

1.7.6 **Addition of amyl alcohol:** Measure 1ml of amyl alcohol and transfer into the butyrometer by means of 1ml pipette or the automatic measure for amyl alcohol.

1.7.7 **Insertion of stopper:** Close the neck of the butyrometer firmly with the stopper without disturbing the contents.

1.7.8 **Mixing of contents:** Shake the butyrometer carefully without inverting it until the contents are thoroughly mixed, the curd is dissolved and no white particles are seen in the liquid. Then invert the butyrometer a few times to mix the contents thoroughly.

1.7.9 Follow the same procedure as described in determination of fat in milk.

1.7.10 **Calculation of percentage of fat** - Multiply the scale reading of the milk butyrometer by the factor 11/10 to obtain the percentage of fat in curd.

1.8 Method for Khoa and Chhana :-

1.8.1 The fat determination of khoa and chhana is same. Three gram of finely grounded homogenous sample of khoa and chhana is weighed into a cheese butyrometer.

1.8.2 Follow the procedures described for method for cheese.

1.8.3 **Calculation of percentage of fat:** Read the percentage of fat by adjusting the fat column with the scale of the butyrometer.

1.9 Method for Ice-Cream :-

1.9.1 Weigh accurately 5 gram of melted sample into the ice-cream butyrometer, then add 6ml of hot water for dilution and washing the sides of butyrometer.

1.9.2 **Addition of acid to butyrometer:** Transfer 10ml of diluted gerber sulphuric acid (70:13 volumes for vanilla ice cream and 94:6 volumes for chocolate ice-cream into the butyrometer).

1.9.3 Follow the procedures described for method for cheese.

Calculation of percentage of fat: Observe the fat column in the butyrometer stem and read the percentage.

Chapter 34

Sterilized Milk Analysis

1. **Determination of Creaming Index**

1.1 **General** - Low creaming index is an indication of good homogenization. Sterilized milk may be graded as under for the quality of homogenization.

Quality of Homogenization	Creaming Index
Excellent	Upto 10
Good	11 to 20
Fair	21 to 30
Bad	Over 30

1.2 **Apparatus :–**

1.2.1 **Glass Tubes :-**

Three, with ground glass stoppers, outside diameter 24mm, length with stoppers 245 mm (suitable for use with a 24 tube Gerber centrifuge), and graduated from 0 to 50ml.

1.2.2 **Pipette :-**

Three, fine pointed, connected to a suitable vessel and to a vacuum pump.

1.2.3 **Apparatus for the Determination of Fat :**

As specified in IS : 1223 - 1958.

1.3 **Procedure :–**

1.3.1 Place 50 ml of the material at 20 $\pm$ 1ºC in each of the three tubes. Centrifuge for 15 minutes at 1000 rev./min.

1.3.2 Using the separate pipette, take 5ml from the upper part of each of the three tubes, carefully taking the cream that adhers to walls of the tubes and transfer into a container (sample-1). Then empty the three tubes into a separate container (sample-II). Measure the fat content in the samples I and II by the Gerber method, taking all the precautions required to be taken for homogenized milk, that is, heating the sample at 65 $\pm$ 2ºC for 5 minutes in a waterbath after each centrifuging before taking the reading till identical readings are obtained. Usually three centrifugings are required each testing for at least 5 minutes.

1.4 Calculation :-

$$\text{Creaming index} = \frac{A - B}{B} \times 100$$

Where,

A = Fat content of the sample I, and

B = Fat content of the sample II.

2. Determination of turbidity:-

2.1 Apparatus :-

2.1.1 Waterbath.

2.1.2 Glass Test Tube - 16 x 160mm size, perfectly transparent.

2.1.3 Erlenmeyer flask - 50ml capacity

2.1.4 Whatman No. 12 or its equivalent filter paper - 12.5cm diameter.

2.2 Reagents :-

2.2.1 Ammonium Sulphate :-

2.3 Procedure :- Place 4 grams of ammonium sulphate and 20 $\pm$ 0.5ml of the sample in a 50ml Erlenmeyer flask at room temperature. Agitate the flask for about one minute until ammonium sulphate dissolves, let the solution stand for at least 5 minutes, then filter it through the pleated filter paper. Collect 5ml of the filtrate in the glass test tube. Place the tube in a boiling waterbath. Examine the tube by moving it before a source of light from which the observers' eyes are screened. The filtrate should be clear.

2.4 Interpretation : Even a slight turbidity would indicate an insufficient heat treatment given to the sterilized milk.

3. Determination of pH.

3.1 Apparatus :-

3.1.1 Incubator - adjusted at 55 ± 1ºC.

3.1.2 pH meter

3.2 Procedure: Determine pH of 50ml of the sample in the flask, with a glass electrode at 20ºC and not the reading. Then incubate another 50ml sample at 55 ± 1ºC for 7 days. Examine the flask each day, then shake and replace it in the incubator. If any physical alteration of the content is observed (coagulation with or without exudation, grittiness, floculation, formation of bubbles or scum, peptonization or proteolysis) the result of the test shall be considered positive and the sample as non-sterile.

If no alteration takes place during the 7 days' incubation at 55ºC, remove the sample from the incubator and cool to room temperature. Take a small portion of it and measure the pH with pH meter with glass electrode at 20ºC, from this pH value subtract the initia' ηH value.

3.3 Interpretation of Results :-

3.3.1 Sample which does not show any physical alteration during incubation at 55 ± 1ºC for 7 days and where the pH does not show a difference of more than 0.3 unit from the initial pH, is considered sterile.

4. Determination of titratable acidity :-

4.1 Apparatus :-

4.1.1 Incubator

4.1.2 Burette - with soda-lime guard tube.

4.1.3 Porcelain Dishes - white hemispherical, of approximately 60ml. capacity.

4.1.4 Stirring Rods - of glass, flattend at one end.

4.2 Reagents :-

4.2.1 Standard Sodium Hydroxide Solution :-

0.1N prepare a concentrated stock solution of sodium hydroxide by dissolving equal parts of sodium hydroxide (sticks or pellets) in equal parts of water in a flask. Tightly stopper the flask with a rubber bung and allow any insoluble sodium carbonate to settle down for 3 to 4 days.

Use the clear supernatant liquid for preparing the standard 0.1N solution. About 8ml of stock solution is required per litre of distilled water. The solution should be accurately standardized against acidic potassium phthalate or oxalic acid.

4.2.2 Phenolphthalein indicator solution :-

Dissolve 1g of Phenolphthalein in 110ml of rectified spirit. Add 0.1N sodium hydroxide solution until one drop gives a faint pink colouration. Dilute with distilled water to 200 ml.

4.2.3 Rosaniline Acetate Stock Solution :-

Dissolve 0.12g of rosaniline acetate in approximately 50ml of rectified spirit containing 0.5ml of glacial acetic acid. Make up to 100ml with rectified spirit.

4.2.3.1. Bench Solution :-

Dilute 1ml of the stock solution to 500ml with a mixture of rectified spirit and distilled water in equal proportions by volume.

Note-

The stock solution and the bench solution should

be stored in dark brown bottles securely stoppered with rubber bungs.

4.3 Procedure :-

4.3.1 Acidity of Fresh Sample :- Weigh 10.0g of the sample into each of two white porcelain basins of approximately 60ml capacity, add to both 10ml of water and stir to disperse the sample. Prepare from one dilution a colour control by adding and stirring 2ml dilute rosaniline acetate solution. Stir 2ml phenolphthalein solution into the other dilution and while stirring vigorously, add as rapidly as possible sodium hydroxide solution from a 10ml burette filled with a sodalime guard tube until the colour matches the pink colour of the control. The titration shall be preferably done in north daylight or under illumination from a daylight lamp.

4.3.2 Acidity After Incubation :- Incubate another 20g of sample at 55 $\pm$ 1ºC for 7 days. Examine the flask each day, then shake and replace it in the incubator. If any physical alteration of the content is observed (coagulation with or without exudation, grittiness, floculation, formation of bubbles or scum, peptonization or proteolysis) the results of the test shall be considered positive and the sample as non-sterile.

4.3.2.1 If no alteration takes place during 7 days' incubation remove the sample from the incubator and cool to room temperature. Weigh 10g of the incubated sample and determine acidity as described in 4.3.1.

4.4 Calculation :-

4.4.1 Acidity of Fresh Sample -

$$\text{Titratable acidity (as lactic acid), per cent by weight} = \frac{9AN}{W}$$

Where,

A = Volume in ml of the standard sodium hydroxide solution required for titration.

N = Normality of the standard sodium hydroxide solution.

W = Weight in gram of the sample taken for the test.

4.4.2 Acidity After Incubation -

Titratable acidity (as lactic acid), per cent by weight = $\frac{9AN}{W}$

Where,

A = Volume in ml of the standard sodium hydroxide solution required for titration.

N = Normality of the standard sodium hydroxide solution.

W = Weight in gram of the sample taken for the test.

4.4.3 Subtract the value obtained in fresh sample of acidity from the value obtained in after incubation of acidity, which would give increase in acidity.

4.5 Interpretation of Results -

4.5.1 Sample which does not show any physical alteration during incubation at 55 $\pm$ 1ºC for 7 days and where the acidity does not show a difference of more than 0.02g from the initial acidity is considered sterile.

5. Alcohol Precipitation Test -

5.1 General :- This is a valuable index of the probable keeping quality of sterilized stored at atmospheric temperatures. It is based on the principle that the mineral balance in sterilized milk may be altered and Rennet is produced due to growth of micro-organisms and this is indicated by the precipitation of proteins when alcohol is added to milk.

5.2 Apparatus :-

5.2.1 Test tubes-

5.3 Reagents :-

5.3.1 80% alcohol (sp. gravity 0.848) containing 0.02% neutral red.

5.3.2 Samples of sterilized milk.

5.4 Procedure :-

5.4.1 Incubate the sample of milk at 37ºC for 24 hours.

5.4.2 Shake the incubated sample thoroughly and transfer 2ml to a test tube.

5.4.3 Add 2ml of the 80% alcohol containing neutral red and shake the contents of the tube gently.

5.4.4 Observe for precipitation or clots.

5.5 Interpretation :- Precipitation or clotting indicates that the keeping quality of the sterilized milk would be less than 21 days when stored at ordinary atmospheric temperature.

6. Determination of total Bacterial Spores :-

6.1 General:- Sterilized milk is expected to be almost completely free from living organisms except for a few highly heat resistant spores surviving the high temperatures to which milk is exposed during the process of sterilization. Sterilized milk should, therefore, keep well for at least 25 to 30 days when stored at ordinary atmospheric temperatures. If the milk is not properly sterilized due to under-heating the number of surviving organisms will be high and they will grow and bring about spoilage of the milk in 2 or 3 days time or even earlier. Leaks in the sealing of the bottles may lead to recontamination of milk.

6.2 Apparatus :-

6.2.1 Dairy Bacteriological Pipettes - to deliver 1ml of milk.

6.2.2 Pipette Container - Preferably of metal, may be round, square or rectangular, length about 400mm.

6.2.3 Petridishes - Outside diameter 98mm, inside diameter about 94mm, depth 15mm with flat bottom and free from bubbles, scratches or other defects.

6.2.4 Petridish Containers - metal boxes with covers, cylindrical or square, for protection and convenient handling of dishes both before and after sterilization.

6.2.5 **Hot Air Oven** - Capable of giving uniform and adequate temperatures, equipped with a thermometer calibrated to read up to 220ºC, and with vents suitably located to ensure prompt and uniform heating.

6.2.6 **Autoclave** - Capable of providing uniform temperatures within the chamber up to the sterilizing temperature of 121ºC, equipped with accurate mercury filled thermometer with bulb properly located so as to register the minimum temperature within the sterilizing chamber (with or without temperature recording instrument) pressure gauge and properly adjusted safety value.

6.2.7 **Incubators** - Two, either water jacketed or an hydric type, electrically heated, thermostatically controlled and provided with shelves so spaced as to ensure uniformity of temperature; one maintained at 30 $\pm$ 1ºC and other 55 $\pm$ 1ºC.

6.2.8 **Hand Tally** - a mechanical counting device.

6.2.9 **Potentiometer or Comparator with Standard Colour Discs** - for accurate determination of pH of media.

6.2.10 **Media Making Utensils** -

Natural, heat resistant glasswares or other suitable non-corrosive equipment, clean and free from foreign residues or foreign material which may contaminate media, such as chlorine, copper, zinc, aluminium, antimony and chromium.

6.3 Reagents :-

6.3.1 Prepare the solid media by dissolving the following ingredients in 1000ml of distilled water.

Ingredients	Weight in grams
Trypton	10
Yeast extract	3
Glucose	1
Soluble starch	1
Agar, bacteriological grade	8

After dissolving the above ingredients, adjust the pH to 7. Place in each 200 x 20mm tubes, about 20ml of media. Sterilize in the autoclave for 15 minutes at 120ºC (or 1kg./cm^2 steam pressure). Media may be stored at 4ºC for not more than 1 month.

6.4 Procedure :-

6.4.1 Pouring Plates - Add aseptically 1ml of the sample to petridishes. Melt the agar medium in the boiling waterbath and cool to 45ºC. Introduce 20ml of the medium aseptically into petridishes within 5 minutes and mix by rotating and tilting the dish without splashing over edge. Spread the mixture evenly over the bottom of the plate. Allow to solidify.

6.4.2 Incubation :- Invert the dishes and incubate at 30 $\pm$ 1ºC and 55 $\pm$ 1ºC separately for 4 days in the incubator.

6.4.3 Counting - Count the colonies with the aid of hand tally.

6.5 **Interpretation :-** The plates should normally show very few colonies (below 100 per ml.) or none at all in the case of sterilized milk samples of satisfactory quality. If the counts are more than 10,000 per ml it is an indication of poor sanitation and gross post-contamination.

Chapter 35

Analysis of Paneer

1. **Definition:-** Paneer is a kind of pressed Chhana and somewhat resembles unripened cheese. According to Indian standards Institution, paneer is an indigenous milk product prepared by the combined action of acid coagulation and heat treatment of cow or buffalo milk or a combination thereof (milk solids suitably processed may be used). The phenomenon of precipitation involves the formation of large structural aggregates of proteins in which milk fat and other colloidal and soluble solids are entrapped along with whey.

2. **Quality Standards for Paneer:-**

   2.1 The prevention of Food Adulteration Rules, 1955.

   Chhana or Paneer means the product obtained from cow or buffalo milk or a combination thereof by precipitation with sour milk, lactic acid or citric acid. It shall not contain more than 70.0 per cent moisture, and the milk fat content shall not be less than 50.0 per cent of the dry matter, milk solids may also be used in preparation of this product.

   2.2 Bureau of Indian Standards:-

   IS: 10484-1983 (Reaffirmed Feb 1999)

Sr. No.	Characteristics	Standard Limits
1.	Moisture % (Max)	60
2.	Fat on dry matter % (Min)	50
3.	Acidity % lactic acid (Max.)	0.5
4.	SPC (Max)	$5x10^5$/g
5.	Coliform (Max.)	90/g
6.	Yeast and mold (Max)	250/g

3. Preparation:-

3.1 Flow Diagram

Milk standardization - 4.5 % fat, 8.5% SNF.

↓

Heating of Milk to 85ºC

↓

Holding At 85ºC for 10 minutes

↓

Cooling to about 78-80ºC

↓

Coagulation at 73-75ºC pH of whey 5.3 to 5.4

↓

Settling of coagulum

↓

Removal of whey

↓

Hooping:- Temperature should be above 63ºC

↓

Pressing at 3-4 Kg/cm^2 for 12-15 minutes

↓

Cooling of Paneer in pasteurized water at 5-10ºC

↓

Brining of paneer (10% salt solution) at 5-10ºC (Optional)

↓

Drying

↓

Packaging

↓

Storage (cold store)

↓

Dispatch (cold chain)

4. Chemical Analysis of Paneer:-

4.1 Determination of Titratable Acidity

4.1.1 Apparatus:-

(i) Burette

(ii) Beaker - 100 ml. capacity.

(iii) Pipette - 10 ml. and 1 ml.

4.1.2 Reagents:-

(i) Standard sodium hydroxide solution.

(ii) Phenolphthalein indicator solution 0.5%.

(iii) Standard hydrochoric acid solution 0.1 N

4.1.3 Procedure:-

(i) Weigh accurately about 2g of paneer into a beaker. Add 3ml of boiling distilled water and mix properly to make fine paste and then add 17 ml. of boiling distilled water, cool to room temperature.

(ii) Add 10ml of sodium hydroxide solution.

(iii) Add 1ml of phenolphthalein indicator solution and mix well.

(iv) Titrate the contents of the beaker, whilst stirring, against the hydrochloric acid solution till the pink colour disappears. Stir vigorously throughout.

(v) Record the volume of hydrochloric acid solution used.

4.1.4 Calculation:-

Titratable acidity (as lactic acid).

Per cent by mass = (10-v)x0.9)/M

Where:

V = volume of hydrochloric acid solution used for titration, and

M= mass in g of paneer.

4.2 Determination of Moisture:-

4.2.1 Apparatus:-

(i) Flat bottom moisture dish with cover 7-8 cm diameter, < 2.5 cm deep.

(ii) Drying oven

(iii) Dessicator

4.2.2 Procedure:-

(i) Weigh accurately about 2 g of the material into a clean dish previously dried and weighed along with a small glass rod.

(ii) Mix the material uniformly with 4ml of hot distilled water with the help of a small glass rod.

(iii) Wash off the particles of material adhering to the glass rod by pouring an additional 1ml of hot distilled water.

(iv) Heat the dish containing the material after uncovering in the oven maintained at $102 \pm 1^{\circ}C$ for about 4 hours.

(v) Cool the dish in the desiccator and weigh with cover on.

(vi) Repeat the process of drying, cooling and weighing at 30 minutes interval until the difference between the two consecutive weighing is less than one mg.

(vii) Record the lowest weight.

4.2.3 Calculation:-

Moisture per cent by mass = 100 (M1-M2)/ (M1-M)

Where,

M1 = mass in g of the dish with material before drying.

M2 = mass in g of the dish with the material after drying, and

M = mass in g of the empty dish.

4.3 Determination of Fat (Mojonnier Method)

4.3.1 Apparatus:-

(i) Fat extraction apparatus.

(ii) Electric Oven - maintained at 100 $\pm$ 1ºC

4.3.2 Reagents:-

(i) Ammonia Solution-approx 25% m/m (Sp. gr. 0.88)

(ii) Ethyl alcohol-distilled Rectified spirit.

(iii) Diethyl ether-Sp.gr. 0.720, free from peroxide.

(iv) Light petroleum ether-boilling range 40 to 60ºC recently distilled.

4.3.3 Procedure:-

(i) Weigh accurately about 1gm of sample into a clean dry 50ml beaker.

(ii) Add 8ml of hot distilled water and then 3ml of ammonia solution. Warm and swirl gently the mixture till paneer is dissolved completely. Cool the mixture.

(iii) With 10ml of ethyl alcohol, transfer the contents to the Mojonnier fat extraction apparatus. Mix well.

(iv) Add 25ml of diethyl ether through the beak or used for weighing the sample. Close the extraction flask and shake vigorously and invert repeatedly for one minute.

(v) Open the flask and add 25ml of light petroleum. Close the flask and shake vigorously and invert repeatedly for one minute. Allow the apparatus to stand until the upper layer has become clear and is distinctly separated from the aqueous layer.

(vi) Carefully transfer as much as possible of the supernatant layer by decantation into a previously dried and weighed flask containing 2 glassbeads. Wash the outside of the neck of the flask and cork with mixed solvent, collecting the rinsing in the flask.

(vii) Make second and third extractions by repeating the procedure using 15ml of each diethyl ether and petroleum ether.

(viii) Collect the solvent in flask and distil carefully the solvent from the flask. Wipe the flask and dry the residual fat in the oven at 100ºC for one hour, taking precaution to remove all traces of volatile solvent and cooling the flask to room temperatures in a desiccator with efficient desiccant and weigh the flask. Repeat heating in oven, cooling and weighing until successive weighing, do not show a loss in mass by more than one mg.

(ix) The difference in mass before and after the petroleum extractions, is the mass of the fat contained in the mass of the Paneer taken.

(x) Calculate the percentage of fat by mass in Paneer.

5. Microbiological Examination of Paneer:-

5.1 Preparation of Diluents

5.1.1 Sodium citrate solution for primary dilution composition

Trisodium citrate, dihydrate	20.0g.
Distilled water	1000 ml.

Preparation:- Dissolve the salt in water (distilled) by heating at 45-50ºC. Adjust the pH so that, after sterilization, it is 7.5 $\pm$ 0.1 at 25 $\pm$ 1ºC.

5.1.2 Diluent for further decimal dilutions-Phosphate buffer solution:-

5.1.3 Composition of stock solution

Potassium dihydrogen phosphate KH2 P04 42.5g

Distilled water 1000 ml.

Preparation:-

Dissolve the salt in 500ml of distilled water. Adjust the pH using 1.0N sodium hydroxide or hydrochloric acid solution so that, after sterilization, it is 7.2 $\pm$ 0.1 at 25ºC $\pm$ 1ºC. Dilute to 1000ml. Sterilize by autoclaving at 121$\pm$ 1ºC or 15 psi for 15 min. Store the stock solution under refrigeration.

5.1.4 Working Solution:-

Add 1 ml of this stock solution (at 20 $\pm$ 1ºC) to 1000ml of water for use as diluent.

5.2 Distribution sterilization and storage of diluents:-

5.2.1 Dispence 90ml of sodium citrate solution into flasks or bottles. Dispence 9ml of working solution of phosphate buffer into test tubes. Sterilize by autoclaving at 121 $\pm$ 1ºC or 15 psi for 15min. The diluents can be stored in dark at a temperature between 0 and 5ºC for one month.

5.2.2 Preparation of Sample and Dilutions:-

(i) Bring the sample of Paneer to room temperature.

(ii) Wipe the exterior of the pack with 70% ethanol to avoid external contamination from the pack.

(iii) Cut open the pack with scissors sterilized by flamming with alcohol. Transfer the contents to a sterile electrical mixer (grinder). Thoroughly grind the contents of the sample.

(iv) Alternately, representative sample of Paneer can be prepared by grinding thoroughly in mortar and pestle sterilized by flaming with alcohol Paneer pieces cut all along the length, width and depth of Paneer block.

(v) Weigh 10g of the prepared sample into the flask containing 90ml of diluent for primary dilution and shake well.

(vi) Transfer 1ml of the primary dilution into a tube containing 9ml of sterile diluent for further decimal dilutions, avoiding contact between pipette and the diluent. A fresh pipette should be used for each dilution.

(vii) Mix carefully, either by aspirating 10 times with a fresh pipette, or in a mechanical shaker for 5-10 seconds to obtain a 10^{-2} dilution.

(viii) If necessary repeat this operation using the 10^{-2} and further dilutions to obtain 10^{-3}, 10^{-4} etc. dilutions until the appropriate number of micro-organisms has been obtained.

5.3 Standard plate count:-

(i) Take two sterile petridishes, using a fresh sterile pipette, transfer to each dish 1ml of the primary dilution of the test sample.

(ii) Take two other sterile petridishes. Using a fresh sterile pipette, transfer to each dish 1ml of the 10^{-2} dilution of the test sample.

(iii) If necessary, repeat the procedure with further dilutions, using a fresh sterile pipette for each decimal dilution.

(iv) Pour about 15ml of the plate count agar medium at $45^{\circ} \pm 0.5^{\circ}C$, in to each petridish.

(v) Carefully mix the inoculum with the medium and allow the mixture to solidify.

(vi) Invert the prepared dishes and incubate them at $37^{\circ}C$ for $48 \pm 2h$.

(vii) Count the colonies in each dish containing not more than 300 colonies.

5.3.1 Expression of Result:-

5.3.1.1 Calculation:-

Calculate the number N of micro-organisms per gram of product.

N= åC/(n 1 + 0.1 n2) d.

Where,

åC is the sum of the colonies counted on all the dishes retained from two successive dilutions and where at least one contains a minimum of 15 colonies.

n1 = the number of dishes retained in the first dilution.

n2 = the number of dishes retained in the second dilution.

d= dilution factor corresponding to the first dilution. Round the result calculated to two significant figures. Express the result as a number between 1.0 and 9.9 multiplied by 10^x where x is the appropriate power of 10.

Example:

No. of colonies at the first dilution (10^{-2}) : 168, 215

No. of colonies at the second dilution (10^{-3}): 14, 25

N = (168+215+14+25)/(2+2x0.1) 10^{-2}

= 422/0.022

= 19181

= 19000 (rounding the result to two significant digits), which can be expressed as $1.9\text{x}10^{4}$ per ml or g of the product.

5.4 Coliform Count:-

(i) Take two sterile petridishes, using a fresh sterile pipette, transfer to each dish 1ml of the primary dilution of the test sample.

(ii) If necessary, repeat the procedure with the further dilutions, using a fresh sterile pipette for each decimal dilution.

(iii) Pour about 15ml of the violet red bile agar (VRBA) medium at 45ºC ± 0.5ºC, into each petridish.

(iv) Carefully mix the inoculum with the medium and allow the mixture to solidify.

(v) After complete solidification, pour about 4ml of the VRBA medium at 45ºC ± 0.5ºC on to the surface of the inoculated medium. Allow solidifying the medium.

(vi) Invert the prepared dishes and incubate them at 37ºC for 24h ± 2h.

(vii) Retain the dishes containing not more than 150 colonies, whether characteristic or not, for counting.

(viii) Count the characteristic colonies in each of the retained dishes.

Characteristic colonies of coliforms are dark or purplish red colonies having a diameter of 0.5 mm or greater and sometimes surrounded by reddish zone of precipitated bile.

5.4.1 Expression of Result:-

5.4.1.1 Calculation:-

Calculate the number N of coliforms per gram of product.

N = åC/(n1 + 0.1 n2) d.

Where:

åC = is the sum of the characteristic colonies counted on all the dishes retained.

n1 = the number of dishes retained in the first dilution.

n2 = the number of dishes retained in the second dilution.

d = dilution factor corresponding to the first dilution.

Round the result calculated to two significant figures. Express the result as a number between 1.0 and 9.9 multiplied by 10^x where x is the appropriate power of 10.

Examples:-

No. of characteristics colonies at the first dilution (10^{-2}) : 83,97.

No. of characteristics colonies at the second dilution (10^{-3})

: 13, 8.

$N = (83+97+13+8)/(2+0.1x2)\ 10^{-2}$

= 9136

Rounding off the result gives 9100, which can be expressed as $9.1x10^3$ per ml or g of the product.

Case: No characteristic colonies.

5.5 Yeast and Mold count:-

(i) Take two sterile petridishes. Using a fresh sterile pipette, transfer to each dish 1ml of the primary dilution of the test sample.

(ii) Take two other sterile petridishes. Using a fresh sterile pipette, transfer to each dish 1ml of the 10^{-2} dilution of the test sample.

(iii) If necessary, repeat the procedure with the further dilutions, using a fresh sterile pipette for each decimal dilution.

(iv) Pour about 20ml. of the chloramphenicol yeast glucose agar medium at 45ºC + 0.5ºC, into each petridish.

(v) Carefully mix the inoculum with the medium and allow the mixture to solidify.

(vi) Incubate the prepared dishes, upright, at 20-25ºC for five days.

(vii) Examine after three days of incubation and count the colonies in those dishes likely to be overgrown before the full incubation period has elapsed.

(viii) Continue incubation for further two days and count the colonies in each dish containing not more than 150 colonies.

5.5.1 Expression of Result:-

5.5.1.1 Calculation:-

Calculate the number N of yeast and molds per gram of product.

N = åC/ (n1 + 0.1n2)d

Where,

åC = is the sum of the characteristic colonies counted on all the dishes retained.

n1 = the number of dishes retained in the first dilution.

n2 = the number of dishes retained in the second dilution.

d = dilution factor corresponding to the first dilution.

Round the result calculated to two significant figures. Express the result as a number between 1.0 and 9.9 multiplied by 10^x where x is the appropriate power of 10

Case: No colonies.

If the two dishes corresponding to the initial suspension, contain no colonies report the result as less than 10 yeast and molds per gram of the sample.

5.6 Media composition and preparation

5.6.1 Plate Count Medium

Tryptone	5.0g
Yeast extract	2.5g
Glucose	1.0g
Agar	15g.
Distilled water	1000ml.
pH (at 25ºC)	7.0 ± 2

5.6.1.1 Preparation:- Dissolve ingredients in 1000ml distilled water. Adjust the pH so that after sterilization, it is 7.0 ± 0.2. Sterilize by autoclaving at 15lbs pressure (121ºC) for 15 minutes. If not used immediately, store in dark at a temperature between 0ºC to 5ºC for no longer than one month..

5.6.2 Violet red bile agar (VRBA) (for coliform count)

Peptic digest of animal tissue	7.0g
Yeast Extract	3.0g
Bile Salts Mixture	1.5g
Lactose	10.0g
Sodium chloride	5.0g
Neutral red	0.03g
Crystal violet	0.002g
Distilled water	1000 ml
Agar	15.0g
pH (at 25ºC)	7.4 ± 0.2

5.6.2.1 **Preparation:-** Dissolve the ingredients in distilled water. Adjust the pH so that after boiling, it is 7.4 ± 0.2 at 25ºC. If necessary. Bring to boil. Allow it to boil for 2 minutes Immediately cool the medium in the waterbath set at 45ºC. Avoid overheating. Consequently do not autoclave the medium. Use the medium within three hours of its preparation.

5.6.3 Chloramphenicol yeast glucose agar (for yeast and mold count)

Yeast extract	5.00g
Dextrose	20.00g
Chloramphenicol	0.100g
Distilled water	1000ml
Agar	15.00g
final pH (at 25ºC)	6.6 ± 0.2

5.6.3.1 Preparation:- Dissolve ingredients in 1000ml distilled water. Adjust the pH so that after sterilization it is 6.6 ± 0.2. Sterilize by autoclaving at 15 lbs pressure (121ºC) for 15 minutes.

Note:- Commercially available dehydrated media may also be used. In such cases, follow the manufactures instructions for preparing the respective media.

Chapter 36

Analysis of Butter

1. **Definition:-** Butter means the product obtained from cow or buffalo milk, or a combination thereof, or from cream or curd obtained from cow or buffalo milk, or a combination thereof, with or without the addition of common salt, and annatto, or carotene as colouring matter. It should contain not less than 80.0% milk fat, not more than 16.0% moisture, not more than 3.0% common salt, and not more than 1.5% of curd. No preservative is permissible in butter.

   Generally table butter is manufactured by churning fresh, pasteurized cream. Butter is also manufactured from ripened neutralised cream.

2. **Quality Standards:-** In India generally following prescribed standards are followed by different organisations;

   2.1 PFA standards.

   2.2 ISI standards

   2.3 Agmark standard

   The above standards are detailed below:-

   2.1 PFA Standards:-

Sr. No.	Characteristics	Standards Limits.
(i)	Fat % (Min)	80.0
(ii)	Curd % (Max)	1.5
(iii)	Salt % (Max)	3.0
(iv)	Diacetyl flavouring agent if used (Max)	4.0 ppm
(vi)	Calcium Hydroxide, Sodium Bicarbonate, Sodium Carbonate and Sodium Polyphosphate to control hydrogen ion concentration (if added) (Max)	0.2
(vi)	Only annatto or carotene is permitted as colouring agent	-

2.2 ISI Standards:

S.No.	Characteristics	Standards Limits
(i)	Coliform organism per ml (Max)	10
(ii)	Yeast and mold per ml	
	Good	Below 20
	Fair	21-50
	Poor	51-100
	Very poor	Over 100

2.3 Agmark Standards:-

Sr.No.	Characteristics	Standards Limits
(i)	Flavour and aroma	Clean, pleasent and free from objectionable taint or rancidity.
(ii)	Body and texture	Firm at 60ºF, should not be greasy or oily. Should be compact and show a uniform granular surface on breaking.
(iii)	Colour	Uniform, should not show streakiness, mottling stain or signs of curd. Only annatto colour is permitted.
(iv)	Moisture % (Max)	16.0 (should be uniformly distributed in the body)
(v)	Fat % (Max)	80.0
(vi)	Acidity % (Max) as Lactic Acid	0.15
(vii)	Salt % (Max)	3.0
(viii)	Curd % (Max)	1.0

3. **Sampling:-** Sampling of butter is done in order to analyse the physical, chemical and bacteriological quality so as to determine its acceptability. It is of great importance to both the producers and consumers as the analysis, based on proper sampling, will indicate if the product corresponds to the prescribed standards

and legal requirements. Sampling should be performed by an authorised and trained staff.

Sampling of butter is carried out at different stages and for both physico-chemical and bacteriological purposes. These have been explained in the following paragraphs.

3.1 Apparatus:-

Following apparatus are required for the proper sampling of butter.

(i) Butter triers of sufficient length to pass diagonally to the base of container.

(ii) Spatulas or Knives

(iii) Wide-mouthed jars - cylindrical and made out of glass, stainless steel or suitable plastic material capable of withstanding, sterilization conditions.

3.2 Sampling methods:-

As stated earlier, sampling of butter is done for different purposes and at different stages of manufacture and storage. These may be for physico-chemical or bacteriological analysis, from bulk, retail pack or from hard and semi-hard butter kept under cold storage etc.

3.2.1 Butter in bulk:-

(i) Prepare dry, clean sampling equipment and containers for samples.

(ii) Insert the trier from surface vertically downwards to the bottom. Make one complete turn and withdraw the full core.

(iii) Hold the point of the trier over the mouth of the sample container and immediately transfer the core of butter from the trier in about 75mm pieces by means of a spatula.

(iv) Take two or more cores of butter so that the minimum mass of total sample is not less than 200gm.

(v) Clean and dry the trier before each insertion.

(vi) The sample jar is hermetically sealed and it is wrapped in paper or stored in dark place until the time of examination/ analysis.

3.2.2 Butter in Packs:-

All the precautions are taken in this case as described above. In addition following steps are taken before actually drawing the sample.

(i) Units weighing 250gm or more are divided in four and two opposite quarters.

(ii) Units weighing less than 250gm are taken as a whole for sampling.

3.2.3 Hard and semi-hard butter kept in cold store:-

While the apparatus and procedure remain same the apportioning of the pieces are done in the following manner:

(i) **From Churns:-** Four cores are drawn at equal distances with the help of a trier. At least two cores are drawn from near the centre of the churn.

(ii) **From trollies:-** Four cores, one each from the two ends and the other from the sides, are drawn with the help of a trier.

(iii) **From boxes:-** Three cores are drawn by inserting a trier vertically through the block while one core would be from the centre, the other two would be drawn from near diagonally opposite corners of the open ends.

(iv) **From casks:-** Three cores shall be drawn by inserting a trier at three points equidistant from the circumference of one end of the block and directed through the centre of the block.

3.3 Sampling for Bacteriological Purpose:

The equipment used for sampling of butter for bacteriological purpose should be thoroughly cleaned, dried and sterilized by one of the following methods:

(i) Immerse the equipment in water at 100ºC for one minute.

(ii) Immerse the equipments in 70% ethanol and put them on flame to burn off the ethanol. In both the cases the sterilization of the equipment should be done immediately before use.

The process of drawing sample remains same as discussed in earlier paragraphs.

4. **Analysis:-** The butter sample is examined for the desirable physical characteristics, analysed for chemical composition and tested for bacteriological quality. All these tests combined together indicate the acceptability of the product. There are various tests in each of the above categories.

4.1 Organoleptic Evaluation:-

The organoleptic evaluation of butter includes the physical examination namely flavour and taste, body and texture, colour, appearance and finish etc.

4.1.1 Flavour and Taste:- The butter sample should be clean, pleasant and free from objectionable taint and bitter taste. In addition, the butter should be free from acid, malty, oxidised, rancid fishy and tallowy flavours.

4.1.2 Body and Texture:-

At 15ºC (60ºF) butter should be firm. It shall not be greasy, oily, leaky, crumbly or sticky. On breaking the butter piece the texture should be found uniform and fine with granular surface.

4.1.3 Others features:-

Colour of butter should be uniform and should not exhibit any streakiness, mottling and stains or signs of curd.

4.1.4 Free Moisture:-

Butter should not exude free moisture on pressing.

Salt:- Salt should be uniformly dissolved.

4.1.5 Packing qualities:-

Packing should be neat, clean and tidy giving good finish.

4.1.6 Score Card for butter:-

During contests and exhibitions a score card is also used wherein, based on organoleptic evaluation 'points' are given to identify the best samples. A score card for butter as designed by ISI is given in table-1.

Table-1
Score Card for Butter (ISI)

Sr.No.	Characteristics	Maximum Points	Minimum Points
(i)	Flavour (clean, free form taint and rancidity)	50	40
(ii)	Body and texture at 15-16ºC (firm, neither greasy nor oily, and showing granular texture on breaking)	20	15
(iii)	Colour, appearance and finish (a) Colour (even i.e. free from streakiness, mottling, stains or signs of curd), (b) Appearance and finish (bright and clean)	20	15
(iv)	Moisture (on pressing, the butter shall not exude beads of free moisture)	10	5
	Total	100	75

4.1.7 Grading of butter is given in table-2.

Table-2
Grading of butter

Score	Quality of butter
95 or above	Excellent
90 to 94	Very good
85 to 89	Good (provided the individual score of each characteristic shall not be less than the minure points mentioned against each).

4.2 Chemical Analysis:-

Chemical analysis plays a great role in determining the quality and acceptability of a product. Butter is analysed for the following contents:

4.2.1 Moisture,

4.2.2 Fat

4.2.3 Salt

4.2.4 Curd

4.2.5 Titratable acidity

4.2.6 Diacetyl & acetyl methylcarbinol

4.2.7 Colouring matters.

In practice the tests for 4.2.1 to 4.2.5 are regularly carried out in a dairy plant. It will be observed that for most of the items there are more than one test available. However, generally at a particular lab one kind of test is conducted unless otherwise required for confirmation.

4.2.1 Determination of Moisture:-

(A) Method-1:-

(a) **Principle:-** The principle of the method is to evaporate the moisture in the butter sample and take the direct reading of percentage from the moisture balance.

(b) Apparatus:

(i) Butter moisture balance.

(ii) Clean and dry aluminium cups/dishes.

(iii) Tong

(iv) Riders.

(v) 10 gm weight

(vi) Spirit lamp or electric hot plate or gas burner.

(vii) Desiccator.

(c) Procedure:-

(i) Adjust the moisture balance to zero with the aid of rider.

(ii) Weigh exactly 10gm of thoroughly mixed sample of butter in a preweighed aluminium cup/dish.

(iii) Hold the cup with tong and with continuous circular motion heat the same over the spirit lamp or burner.

(iv) Continue heating until the foam, produced during heating, has ceased and the curd at the bottom of the cup has attained the characteristic of slightly brown (golden) colour.

(v) Allow the cup to cool and place it in the desiccator.

(vi) Place the cooled dish in the moisture balance and read the percentage directly.

(B) Method-II

(a) Principle:- The principle of this method is to evaporate the moisture from the sample, determine the loss of moisture and calculate its percentage against the original weight of the sample.

(b) Apparatus:-

(i) Drying oven

(ii) Flat bottom moisture dish.

(iii) Glass rod

(iv) Dessiccator

(v) Waterbath

(vi) Clay pipe triangle

(c) Procedure:-

(i) Accurately weigh about 3-4 gm of well mixed sample of butter in a clean, dry flat bottom moisture dish and record its net weight (x).

(ii) Place the dish on a steambath, supported on a clay pipe triangle, for at least 20 minutes.

(iii) Stir at frequent intervals until no moisture is seen at the bottom of the dish.

(iv) Remove the dish, wipe off the bottom of the dish and transfer it to the oven maintained at 100ºC $\pm$ 1ºC and hold it for 90 minutes.

(v) Cool the dish in the desiccator and weigh.

(vi) Repeat the process of heating, cooling and weighing until the difference between two consecutive weights does not exceed 0.1 mg.

(vii) Record the net weight of the residual matter on the dish (y)

(d) Calculation:- Moisture % by weight = X-Y x 100/ X

4.2.2 Determination of fat Content:-

Determination of fat percentage in butter is performed on the same principle as for milk and cream. For all practical purposes fat is determined by Gerber Method; although other methods are also prescribed.

(A) Method-1

(a) Apparatus:-

(i) Analytical balance

(ii) Porcelain dish with stirring rod.

(iii) Butyrometer for butter fat-graduated 70%-90%.

(iv) Water bath

(v) Centrifuge.

(b) Reagents:-

(i) Sulphuric acid-density 1.50 gm/ml at 15.5ºC.

(ii) Amyl Alcohol-density 0.814–0.816 gm/ml at 15.5ºC.

(c) Procedure:-

(i) Mix the butter sample in a porcelain dish thoroughly.

(ii) Weigh accurately 5 gm of butter in the cup of the butyrometer and replace the cup in the butyrometer.

(iii) Fill 20ml sulphuric acid and then 1 ml amyl alcohol in the butyrometer.

(iv) Fix stopper and shake the butyrometer until the proteins are dissolved.

(v) If the contents in the butyrometer does not reach the 70% mark, add required amount of sulphuric acid to reach the mark.

(vi) Place the butyrometer in waterbath at 50ºC- 53ºC for 4-5 minutes to enable melting and separation of the fat.

(vii) Centrifuge for 3-4 minutes.

(viii) Take the butyrometer from the centrifuge and place it in waterbath for 5 minutes.

(ix) Take the butyrometer from the waterbath and read the amount of fat in per cent.

(B) Method-II (ISI Method)

(a) Apparatus:

(i) Flat bottom dish

(ii) Waterbath

(iii) Hot air oven

(iv) Desiccator

(v) Filter paper

(vi) Flask

(vii) Hot plate

(b) Reagents:-

(i) Petroleum ether (boiling point 40-60ºC)

(c) Procedure:-

(i) Weigh about 10 gm of butter into a previously cleaned and dried flat bottom dish and record the net weight of the content (x).

(ii) Place the dish on steam bath to remove the moisture as discussed in method-II.

(iii) Put the dish on the hot air oven for 90 minutes at 100ºC $\pm$ 1ºC.

(iv) Cool the dish in the desiccator.

(v) Add about 25-50ml of petroleum ether to the dish and mix well.

(vi) Filter the solution (in cash of decanting, care should be taken to leave the sediment undisturbed).

(vii) Collect the filtrate (ether extraction) in a 250ml flat bottom flask containing glass beads which is previously cleaned, dried and tared.

(viii) Wash all the fat and sediment from the dish with 20-25ml of petroleum ether and finally wash the filter paper with petroleum ether until it is free from fat.

(ix) Collect all the filtrate in the flask.

(x) Evaporate the petroleum solvent from the filtrate by putting the flask on the hot plate.

(xi) Dry the flask, containing the fat, in an oven for one hour at 100ºC ± 1ºC.

(xii) Cool the flask in a desiccator and weigh until the loss of weight between consecutive weighing does not exceed 0.1mg.

(xiii) Record the net weight of the content (Y)

(d) **Calculation:-** Fat % by weight = Y/ X x 100

4.2.3 Determination of Salt:-

(A) **Method-I:-**

The determination of salt content in butter is based on the reaction of sodium chloride and silver nitrate in definite proportion. By titrating with silver nitrate, using potassium chromate as an indicator, the end point is denoted by formation of an orange brown precipitate.

(a) **Apparatus:-**

(i) Analytical balance (ii) 25ml volumetric pipette. (iii) 250ml Volumetric flask. (iv) Burette graduated to 0.1ml (v) Glass rod (vi) Porcelain dish.

(b) **Reagents:-** (i) Silver nitrate solution 0.1 N. (ii) Potassium chromate solution 5% (m/v) in distilled water (iii) Distilled water.

(c) **Procedure:-**

(i) Make a blank test using the reagents in the same amount, without butter sample, as per procedure described below and record the volume of silver nitrate used (vo).

(ii) Weigh accurately about 5g of the well mixed sample of butter in the conical flask (w).

(iii) Carefully add about 100ml of boiling distilled water, allow proper mixing and cool to 50-55ºC

(iv) Add 2ml of potassium chromate solution and mix by gentle swirling.

(v) Titrate the content with silver nitrate solution (0.1N) until the colour change to orange brown persists for 30 seconds. Record the volume of silver nitrate used (vi)

(d) Calculation:-

Salt % = 5.85 x N x (V1 - VO) / W

(exposed as % by mass) of NaCl

Where N = Normality of silver nitrate solution.

Note:- The difference between results of duplicate determination (results obtained simultaneously or in rapid succession by the same analyst) should not exceed 0.02g sodium chloride for 100g of the product.

This is a direct method as prescribed under ISI.

(B) Method-II (ISI Routine Method)

In this method salt is extracted with hot distilled water from dried fat fat-free residue obtained after moisture and fat determination. The chloride is precipitated by the addition of excess silver nitrate solution. The unused silver nitrate is back titrated with potassium thiocyanate using ferric ammonium sulphate as indicator.

(a) **Apparatus:-** (i) Volumetric flask. (ii) Conical flask-250 ml. (iii) Burette (iv) Pipette.

(c) Reagents:-

(i) Ferric ammonium sulphate indicator Solution-Dissolve 50g of ferric ammonium sulphate in 95 ml of water containing 5ml of 5 N nitric acid.

(ii) Standard silver nitrate solution 0.05 N.

(iii) Standard potassium thiocyanate solution. - 0.05 standardized against silver nitrate.

(iv) Nitric acid-approximately 5 N.

(v) Nitrobenzene.

(vi) Nitric acid-Sp. gr. 1.42 (approximately 70% w.w).

(c) Procedure:-

(i) Extract the salt from the residue of curd and salt after the determination of moisture and fat using hot water. Collect the rinse also in 100ml flask.

(ii) Cool to room temperature and make upto volume.

(iii) Take 25ml water extract into a 250ml conical flask.

(iv) Add an excess (normally 25 to 30 ml.) of 0.05 N silver nitrate solution.

(v) Acidify with nitric acid and add 2ml of the indicator solution and 1ml of nitrobenzene and mix.

(vi) Determine the excess of silver nitrate by titration with the potassium thiocyanate until the appearance of an orange tint which persists for 15 seconds.

(vii) In the same manner determine the equivalent of 25ml or the added amount of the silver nitrate as thiocyanate, using the same volumes of reagents and water.

(d) Calculation:- Salt (Nacl) % by weight = 23.38 x N x (A-B)/ W

Where: N = Normality of potassium thiocyanate.

A = Volume of potassium thiocyanate in the blank titration.

B= Volume of potassium thiocyanate in the sample titration.

W= Weight of the sample in gms.

4.2.4 Determination of curd.

Normally butter should not have more than 1.5% curd. It is estimated by making fat-free the residue obtained after the determination of moisture. The residue is dried and weighed.

(a) **Apparatus:-**

(i) Gooch crucible or sintered funnel with filter flask and adapter.

(ii) Glass funnel (iii) Flat bottom flask-250 ml (iv) Desiccator (v) Asbestos (vi) Whatman or equivalent filter paper- 12.5cm.

(b) **Reagents:-** (i) Petroleum hydrocarbon solvent-boiling range 40-60ºC.

(c) **Procedure:-**

(i) Prepare an asbestos mat in a Gooch crucible or sintered funnel and dry it in an oven at 100ºC to $\pm$ 1ºC and weigh.

Alternatively, dry cool and weigh ordinary glass funnel with folder 12.5 cm filter paper. Record the weight(x).

(ii) Melt the residue in the moisture dish or cup obtained from the moisture determination.

(iii) Add 25 to 30ml of petroleum solvent and mix thoroughly.

(iv) Fit the crucible to the filter flask or place the funnel with filter paper on a filter stand.

(v) Wet the asbestos mat or the filter paper with petroleum solvent and decant the fatty solution from the dish into the asbestos or filter paper leaving the sediment in the dish.

(vi) Macerate the sediment twice with 20-25ml of petroleum solvent and decant the fatty solution again into the asbestos or the filter paper.

(vii) Filter the solution and collect the filtrate in a clean, dried, tared 250 ml flat bottom flask containing a glass bead.

(viii) With the aid of a wash bottle, containing petroleum solvent, wash all the fat and sediment from the dish into the crucible or the filter paper.

(ix) Finally, wash the crucible or the filter paper until free from fat.

(x) Dry the crucible or filter funnel in the oven maintained at 100ºC $\pm$ 1ºC for at least 30 minutes.

(xi) Cool in the desiccator and weight.

(xii) Repeat drying, cooling and weighing until the loss of weight between the consecutive weighings does not exceed 0.1mg. Record the final weight (y).

(d) **Calculation** :- Curd + Salt % by Weight = 100 (Y-X)/ W

Where W= Weight in gms of sample.

For determining the % of curd, analyse the salt % and deduct the latter from curd + salt %.

4.2.5 Determination of titratable Acidity:-

Butter is melted by hot water and the hot solution is titrated with standard alkali till neutral to phenolphthalein.

(a) **Apparatus:-** (i) Burette, (ii) Conical flask,

(b) **Reagent:-** (i) Standard NaoH Solution- 0.02 N.

(ii) Phenolphthalein indicator solution.

(c) **Procedure:-** (i) Weigh accurately 20g butter sample in a 250ml dry conical flask.

(ii) Add 90ml hot boiled water.

(iii) Shake the content well.

(iv) Add 1ml phenolphthalein indicator.

(v) Titrate the hot sample with 0.02 N NaoH and take the burette reading.

(d) Calculation:- Titratable acidity % by weight = NxVx9/ W

(as lactic acid)

Where : N = Normality of NaoH Solution.

V = Volume of NaoH used.

W= Weight in gms of the sample.

4.2.6 Determination of diacetyl and acetylmethyl carbinol (ISI Method):-

The diacetyl in butter is removed by distillation and is estimated. For acetylmethyl carbinol (AMC), a separate sample is first treated with ferric chloride which oxidises AMC to diacetyl and the latter is then distilled and estimated.

(a) Apparatus:-

(i) Distillation flask - 2 litres capacity.

(ii) Beakers - 200ml capacity.

(iii) Sintered glass crucibles - with filter flask and adopter.

(iv) Desiccator-with efficient desiccant

(v) Oven – maintained at 120ºC

(b) Reagents:-

(i) Hydroxylamine hydrochloride (AR) - 20% soln.

(ii) Sodium acetate - 20% solution.

(iii) Nickel chloride – 10% solution

(iv) Ferric chloride - 40% solution

(v) Mixed reagent:- Mix 4ml each of 20% solutions

of hydroxylamine hydrochloride and 20% sodium acetate plus 2ml 10% nickel chloride solution. Filter and keep the mixed reagent in a glass stoppered bottle protected from light (prepare fresh mixed agent every time).

(c) **Procedure for diacetyl:-**

(i) Weigh 400 gms of butter into a two litre distillation flask and steam and distil into 10 ml of the freshly prepared mixed reagent.

(ii) Continue steam distillation untill 100ml distillates has been collected.

(iii) The glass tube at the end of the container must be submerged in the reagent.

(iv) After completing the distillation, heat the distillate at 60ºC and allow to stand overnight.

(v) Filter into a weighed dried Gooch crucible.

(vi) Set the filtrate aside for another 24 hours and again filter out any precipitate that may have formed.

(vii) Dry the precipitate for 2 hours at 120ºC with not less than 50mm of vacuum.

(d) **Result:-** Weigh and express the result as mg of nickel salt (nickel-dimethyl glyoxime) per kg of butter.

(e) **Procedure (for acetylmethyl carbinol):-**

(i) Add 40ml of filtered solution of ferric chloride (40%) to 400 gms of butter in a distillation flask.

(ii) Proceed as in diacetyl determination.

(f) **Result:-** The results of this distillation will give the combined value of diacetyl and AMC contents. This is preferred method for analysing butter as only one distillation is required. To determine the amount of AMC present, subtract the result of diacetyl determination from that for AMC. Express the result as mg of nickel salt per kg of butter.

4.2.7 Detection of colouring matters:-

(a) Reagents:-

(i) Ethyl ether, (ii) Hydrochloric acid solution 1:1 (iii) Sodium hydroxide solution - 10% (iv) Sodium hydroxide solution - 20% (v) Stannous chloride solution - 40%, containing sufficient concentrated hydrochloric acid to make the solution acidic and a small piece of tin to keep it reduce. (vi) Filter paper.

(b) Procedure:-

(i) Dissolve 5 grams of butter in ethyl ether, filter, and collect the ether fraction.

(ii) Pour about 2 ml each into two test tubes.

(iii) To one test tube, add 1 to 2ml of 1:1 solution of hydrochloric acid.

(iv) To the other test tube add the same volume of 10% sodium hydroxide solution. Shake the tubes well and allow them to stand.

(v) In the presence of some azodyes acid solution shows pink to win-red colour, while the alkaline solution in the other tube shows no colour.

(vi) If, on the other hand, annatto or other vegetable colour is present, alkali solution is coloured yellow, while no colour appears in acid solution.

(vii) Red colour changing to yellow, especially on warming, in alkaline solution, may be due to presence of gallate anti-oxidants.

(viii) Pour on moistened filter paper alkaline solution of colour obtained by shaking clear butter with warm 2% sodium hydroxide solution.

(ix) If annato is present, paper absorbs colour, so that

when washed with a gentle stream of water it remains dyed straw colour.

(x) Dry the filter, add one drop of 40% stannous chloride solution.

(xi) Again dry carefully.

(xii) If colour turns purple, presence of annatto is confirmed.

4.3 Bacteriological Analysis of Butter

Bacteriological examination of any food product is of utmost importance to the consumers and butter is no exception. It is more important for butter because of its easy perishable nature due to presence of moisture and other nutrients capable of accelerating the growth of micro-organisms namely bacteria, yeast and mold etc. both spoilage and/or pathogenic. During the manufacture of butter, cream is subjected to come in contact with human being, pipelines and machineries and additives which are generally the possible contaminating agents.

4.3.1 Standard Plate Count (SPC):-

(a) Apparatus:-

(i) Petridish - sterile.

(ii) 1 ml pipette - sterile

(iii) Bunsen burner

(iv) Waterbath

(v) Colony counter

(b) Reagent:- Sterilized agar medium.

(c) Procedure:-

(i) Warm and melt the sample of butter in a sterilized container in waterbath maintained at 42.5ºC for a period not exceeding 15 minutes.

(ii) Properly agitate the container to get uniform mixing of serum, water and fat.

(iii) Warm the sterile dilution blanks at 40ºC in a waterbath.

(iv) Draw 1ml of melted butter by 1ml pipette and pour in 99ml of dilution blank taking all precautions. This will give 1:100 dilution of the sample. Mix properly.

(v) Pipette out 1ml of this into sterile petridish in duplicate and add about 15ml sterilized agar into it. The pouring should be done carefully exposing the mouth of the media flask to the flame.

(vi) Mix the content by gently rotating and tilting without spreading over the edges of the petridishes.

(vii) Allow the media to set, invert and keep in an incubator at 37ºC $\pm$ 0.5ºc for 48 hours.

(d) **Results:-** By the colony counter count number of colonies developed, compute the numbers of colonies per multiplied by the dilution factor. In this case due allowance should be given for other number of colonies developed in the control plate. Express the result as number of colonies per gm of butter.

4.3.2 Coliform Count:- Butter, manufactured from pasteurized cream, should not have any coliform. The presence of coliform indicates contamination during the process of manufacture.

(a) **Apparatus:-**

(i) Pipette (Sterile)- 10ml and 1ml

(ii) Petridish (Sterile)

(iii) Waterbath- at 43ºC to 45ºC

(iv) Incubator - at 37ºC $\pm$ 0.5º

(v) Volumetric flask.

(b) **Reagent:-** (i) Dilution blank

(ii) Violet red bile agar.

(iii) Desoxycholate (lactose) agar.

(c) Procedure:-

(i) Warm and melt the test sample of butter in the sterile container by keeping it in a waterbath maintained at 43ºC to 45ºC for a period of not exceeding 15 minutes.

(ii) Agitate thoroughly so as to obtain uniform mixing of the serum, water and fat and, if necessary, use a sterile glass rod for complete mixing.

(iii) Warm dilution bank to about 40ºC in the waterbath and sterile 10ml and 1ml pipettes by drawing in and forcing out the warm dilution blank for 2 or 3 times.

(iv) Transfer 10ml of the melted butter into the sterile dilution blank (warm to 40ºC) in a sterile 100ml volumetric flask and make up the volume to give a dilution of 1:10.

(v) Transfer 1ml of this dilution to a 9ml blank in a sterile flask to obtain 1:100 dilution.

(vi) Transfer 1ml of this solution into sterile petridishes in duplicate.

(vii) Add to each petridish 10 to 15ml of violet red bile agar or desoxychloate agar previously melted and cooled to 42-44ºC.

(viii) Mix the contents thoroughly by gentle tilting and rotation of the plates.

(ix) After the mixture has solidified, pour another layer of the same medium (5 to 6ml and spread evenly to cover the surface completely.

(x) When the medium has set, invert and incubate the petridishes at 37ºC $\pm$ 0.5ºC for 24 hours.

(xi) Examine the plates for presence of typical colonies of coliform bacteria indicated by dark red colonies measuring at least 0.5mm in diameter.

(xii) Count all such colonies and report the result as number of colonies per millilitre of butter.

(d) **Results:-** Coliform counts exceeding 10 per ml of butter shall be taken to indicate inefficient pasteurization of cream or contamination of the product from wash water, equipment and other sources during manufacture and packing.

4.3.3 Determination of yeast and mould count:-

Total bacterial count cannot logically be used in determining the general condition surrounding the manufacture and handling of butter because culture of specified organisms are frequently added to the cream and occasionally directly to the butter itself, with the result that the bacterial content of the finished butter is influenced. Yeast and mould counts of butter have, accordingly, been suggested because these micro-organisms should be present, if at all, in very small number.

(a) **Apparatus:-**

(i) Sterile, screw-cap or glass-stoppered glass bottle of suitable sizes.

(ii) Petridishes - sterile, with cover.

(iii) Pipettes - sterile 1.1ml, 10ml and 11ml.

(iv) Waterbath – maintained at 43 to 45ºC

(v) Incubator - maintained at 25ºC $\pm$ 1ºC

(b) **Reagents:-** (i) Potato glucose agar (acidified).

(c) **Procedure:-**

(i) Warm the sample of butter contained in the sterile jar, as well as sterile buffered water blank, to about 40ºC in a waterbath maintained at 43ºC to 45ºC (the time required for melting the butter should not exceed 15 minutes).

(ii) Agitate thoroughly so as to obtain uniform mixing of the serum, water and fat.

(iii) With a previously warmed sterile 10ml pipette,

transfer 10ml of butter to a 90ml sterile buffered water blank, which is at 34 to 40ºC (11ml of butter to 99ml. of buffered water to give the same 1:10 dilution)

(iv) Shake this dilution for 25 times in the usual manner just before inoculating the petridishes with the different dilutions given below in duplicate.

(a) 1:2 (5ml of the 1:10 dilution)

(b) 1:10 (1ml of the 1:10 dilution)

(c) 1:100 (0.1ml of the 1:10 dilution)

(v) Prior to pouring, adjust reaction in each container to pH 3.5±0.1. Because remelting of acidified medium may destroy its solidifying properties, adjust the amount needed for immediate plating.

(vi) The petridishes containing different dilutions are floaded with the melted adjusted potatodextrose agar.

(vii) After solidification, the plates are incubated for 5 days at 21ºC to 25ºC.

(d) **Results:-** To give the actual colony counts per millilitre of butter, the colony counts obtained from:

(i) 1:2 dilution should be multiplied by 2

(ii) 1:10 dilution by factor 10

(iii) 1:100 dilution by factor 100.

Recommended standard total yeast and mould count per ml of butter.	Sanitary index.
below 20	Good
21 to 50	Fair
51 to100	Poor
Over 100	Very poor.

(e) **Interpretation:-** High yeast and mould count estimates in freshly churned samples indicate.

(i) Ineffective cleaning and sterilizing.

(ii) Inefficient pasteurization

(iii) Carelessness in cleaning and handling the equipment.

Note:- For more details please refer to IS:3507 - 1966.

Chapter 37

Ghee Analysis

1. **Determination of Moisture Content**

1.1 Apparatus:-

1.1.1 Moisture Dish – of aluminium, 7-8 cm in diameter and 2-2.5cm deep, provided with tight fitting slip over cover.

1.1.2 Deciccator – Containing an efficient desiccant, such as phosphorus pentoxide.

1.1.3 Air-oven:-

Preferably electrically heated, with temperature control device.

1.2 Procedure :-

Weigh accurately about 10g of the sample into a moisture dish which has been dried previously, cooled in the desiccator and then weighed. Place the dish in the air-oven for approximately 1 hour at 105 $\pm$ 1ºC. Remove the dish from the oven, cool in the desiccator to room temperature and weigh. Repeat this procedure but keep the dish in the oven only for half an hour each time until the difference between the two successive weighings does not exceed 1mg. Preserve the dried sample for the determination of insoluble impurities.

1.3 Calculation – Moisture and volatile matter content per cent by weight

$$= \frac{100\ (w1 - w2)}{(w1 - w2)}$$

Where,

W1 = Weight in g of the dish with ghee before drying.

W2 = Weight in g of the dish with ghee after drying, and

W = Weight in g of the empty dish.

1.4 Accuracy of the Method - The maximum deviation between duplicate determinations shall not exceed 0.1 (per cent).

2. **Determinations of Colour** - Two methods for measuring colour of ghee are prescribed. The first method using a Tintometer is simple and suitable for routine work. Where more precise information is required the second method using a spectrophotometer shall be used.

2.1 Tintometric Method :-

2.1.1 Apparatus

2.1.1.1 Tintometer - preferably with light attachment.

2.1.1.2 Thermometer - calibrated from 0-50ºC

2.1.1.3 Tintometer cells - 0.5cm and 1cm.

2.1.1.4 Waterbath - maintained at 40-50ºC.

2.1.2 Procedure :- Melt the sample and transfer it to the Tintometer cell. Keep the cell in a waterbath and stir the contents with a thermometer. When the sample attains a temperature of 40ºC, match the colour against standard glasses in the Tintometer. Express the results as yellow units per cm at 40ºC.

2.2 Spectrophotometer Method :-

2.2.1 Apparatus

2.2.1.1 Spectrophotometer - A spectrophotometer capable of adjustment to give the following readings on a standard nickel sulphate solution at 25-30ºC, after setting the zero point and after adjusting the 100 per cent transmittance point (of absorbance) against carbon tetrachloride in a cuvette.

Millimicrons	Transmittance
400	Less than 4.0 per cent
460	26.2 ± 2.0 per cent
510	73.9 ± 1.0 per cent
550	54.8 ± 1.0 per cent
620	5.2 ± 0.5 per cent
670	1.1 ± 0.5 per cent
700	Less than 2.0 per cent

2.2.1.2 Matched glass cylindrical cuvettes- Inside diameter approximately 21.8mm, outside diameter approximately 24.5mm. All cuvettes to be used with a given instrument should be checked with carbon tetrachloride (CCl4) and nickel sulphate solution at 550nm and give within ± 0.6 per cent of the same transmittance. The cuvettes should be kept clean and free from scratches.

2.2.1.3 Standardizing nickel sulphate solution :-

Dissolve 200 gram of nickel sulphate (Niso4. 6H20), analytical reagent grade, in distilled water. Add 10ml of concentrated hydrochloric acid. Dilute to exactly 1000ml in a volumetric flask. The temperature of the solution should be between 25ºC and 30ºC. The nickel content of the solution shall be between 4.40 and 4.46g of nickel per 100ml at 25-30ºc.

2.2.1.4 Filter paper - Fine porosity, such as Whatman No.12

2.2.2 Reagents :-

2.2.2.1 Carbon tetrachloride - redistilled if the transmittance differs from distilled water by 0.5 per cent at 400mμ.

2.2.3 Procedure :-

2.2.3.1 Calibration of the Spectrophotometer -

Turn on the spectrophotometer and allow at least 20 minutes warm-up period before standardizing or making any measurements. After the initial warm-up period, rotate both the control knobs on top of the instrument counter-clockwise to their stop position. Adjust the galvanometer by means of the galvanometer adjustment or by sliding the scale so that an exact zero reading is obtained. Set the wavelength dial to 460mμ. Recheck the zero reading of the instrument, insert a cavette filled with carbon tetrachloride in the instrument and set the 100 per cent transmittance point exactly. Fill a cuvette with the standardizing nickel sulphate solution and read the transmittance of the solution. The reading should fall between 24.2 and 28.2. If the reading at 460mμ is above 26.2, the reading at 500mμ should be above 54.8, if the reading at 460mμ is belolw 26.2, the reading at 550mμ. should be below 54.8, otherwise adjust the wavelength knob underneath the instrument until both reading are in the same direction above or below the median values established, but within the specific limits.

Set the instrument at 510mμ and read the transmittance for the nickel sulphates solution. The 510mμ reading shall be between 72.9 and 74.9. Read the other specified values. All should fall within the limits specified.

2.2.3.2 **Determination:-** The sample shall be rendered optically clear and free from water and other suspended impurities. Adjust the temperature of the sample to 35-40ºC, fill the cuvette using a sufficient amount of sample to ensure a full column in the light beam. Place the filled cuvette in the instrument and read the absorbance to the nearest 0.00/ at 460, 550, 620 and 670mμ.

2.2.4 Calculation - Photometric Colour = 1.29 D460 + 69.7 D550 + 41.2 D620 - 56.4 D670

Where D is the absorbance.

2.2.4.1- Special instrument scales for reading the four factors involved directly may be used.

3. Determination of Refractive Index :-

3.1 Refractive index is the ratio of the velocity of light in vacuum to the velocity of light in the sample medium, more generally, it is expressed as the ratio between the sine of the angle of incidence to the sine of the angle of refraction when a ray of light of a definite known wavelength (usually 589.3mμ. the mean of the D-lines of sodium) passes from air into ghee. Refractive index of ghee is measured at 40ºC to ensure that the sample is completely melted.

3.1.1 Accurate results are obtained by using monochromatic light of a wavelength of 589.3mμ., (the mean of the D-lines of sodium). Diffused white light may be used provided the instrument used is fitted with a suitable compensator. Readings with white light are only accurate when a perfectly colourless and sharp line of demarcation is obtained between the dark and light shades.

3.1.2 The refractive index should be read on an Abbe refractometer which gives the true refractive index or on a butyro-refractometer, which reads on an arbitrary scale at constant temperature as near 40ºC as possible.

3.2 Apparatus :-

3.2.1 Precision Refractometer - fitted with an accurate thermometer (reading from 40 to 50ºC)

3.2.2 Hot water circulating device - to maintain the temperature of the prism constant at 40 $\pm$ 1ºC.

3.2.3 Sodium Lamp - Daylight can also be used if the refractometer has an achromatic compensator.

3.2.4 Sodium Lamp - Daylight can also be used if the refractometer has an achromatic compensator.

3.3 Reagent - Standard fluid for checking the accuracy of the instrument.

3.4 Procedure :-

3.4.1 The sample shall be rendered optically clear, and free from water and other suspended impurities.

3.4.2 The correctness of the instrument shall be tested before taking the reading by carrying out tests with fluid of known refractive index. At temperature of 40ºC or over, the prisms of most instruments never reach the temperature indicated by the registering thermometer and at temperatures greatly removed from the standard temperature for the instruments, there is a small error due to the change of the refractive index of the glass. At these high temperatures check the instruments experimentally with a liquid of known temperature co-efficient, and apply the correction thus found to instrument readings given by the sample.

3.4.3 It shall be borne in mind that the presence of free fatty acids considerably lowers the refractive index.

3.4.4 Ghee shall completely fill the space between the two prisms and shall show no air bubbles. The readings shall be taken after ghee has been kept in the prisms for 2 to 5 minutes and after it has been ensured that it has attained constant temperature by taking two or more readings. Take care that the ghee has reached the temperature of the instrument before the readings is taken. Before commencing to take readings circulate through prisms a stream of water at constant temperature and measure accurately the constant temperature at which the readings are taken.

3.4.5 Use of Abbe Refractometer - To charge the instrument, open double prisms by means of screw head and place a few drops of the sample on prisms or, if preferred, open prisms slightly by turning screw head and put a few drops of sample into a funnel-shaped aperture between prisms. Close prisms firmly by tightening the screw head. Allow instrument to stand for a few minutes before reading is taken so that temperature of the sample and instrument will be the same.

3.4.5.1 Method of measurement is based upon observation of position of border line of total reflection in relation to the faces of a prism of flint glass. Bring this border line into field of vision of telescope by rotating the double prism by means of the alidade in the following manner.

Hold sector firmly and move alidade backward or forward until field of vision is divided into light and dark portion. Line dividing these portions the "border line" and, as a rule, will not be a sharp line but a band of colour. The colours are eliminated by rotating screw head of compensator until sharp, colourless line is obtained. Adjust border line so that it falls on the point of intersection of cross-hairs. Read refractive index of the substance directly on the scale of sector. Check correctness of instrument with water at 20ºC, the theoretical refractive index of water at 20ºC is 1.3330. Any correction found necessary should be made on all readings.

Maximum difference between duplicate determinations shall not exceed 0.0002 unit of the refractive index.

3.4.6 Use of Butyro-Refractometer -

Place 2 or 3 drops of the sample on surface of

lower prism. Close the prism and adjust as in 3.4.5.

3.4.7 For conversion of refractive index values into butyro-refractometer reading and vice versa, use Table -1

Table -1
Butyro-Refractometer Readings and Indices of Refraction.

B.R. Reading	Refractive Index
(1)	(2)
35.0	1.4488
35.5	1.4491
36.0	1.4495
36.5	1.4499
37.0	1.4502
37.5	1.4506
38.0	1.4509
38.5	1.4513
39.0	1.4517
39.5	1.4520
40.0	1.4524
40.5	1.4527
41.0	1.4531
41.5	1.4534
42	1.4538
42.5	1.4541
43.0	1.4545
43.5	1.4548
44	1.4552

44.5	1.4555
45	1.4558
45.5	1.4562
46.0	1.4565
46.5	1.4569
47	1.4572
47.5	1.4576
48	1.4579
48.5	1.4583
49	1.4586
49.5	1.4590
50	1.4593

3.4.8 The refractive index decreases with a rise and increases with a fall in temperature. If the temperature is not exactly at 40ºC, X is added to the observed reading for each degree above or subtracted for each degree below 40ºC prorata, where

X for butyro - refractometer = 0.55

X for Abbe refractometer = 0.000365

Normally the temperature of observation shall not deviate by more than $\pm$ 2ºC.

3.4.9 Accuracy of the Method :-

The maximum difference between duplicate determinations shall not exceed 0.0002 unit for the refractive index and 0.1 for the butyro-refractometer reading.

4. Determination of Acidity :-

4.1 Apparatus :-

4.1.1 Conical Flasks - 250ml capacity.

4.1.2 Burette - with soda lime guard tube.

4.2 Reagents :-

4.2.1 Ethyl Alcohol or Rectified Spirit - 95 per cent (v/v), sp. gr. 0.8160 neutral to phenolphthalein.

4.2.2 Sodium Hydroxide or Potassium Hydroxide - 0.1 N agueous solution accurately standardized against acid potassium phthalate (AR) or oxalic acid (AR).

4.2.3 Phenolphthalein Indicator -1.0 per cent solution in 95 per cent (v/v) ethyl alcohol or rectified spirit.

4.3 Procedure :-

Weigh 10g of the sample in a 250ml conical flask. In a second flask bring 50ml of alcohol to the boiling point and while still above 70ºC neutralize it to phenolphthalein (using 0.5ml) with 0.1N sodium hydroxide. Pour the neutralized alcohol on ghee in the flask and mix the contents of the flask. Bring them to boil and while it is still hot, titrate with 0.1N sodium hydroxide, shaking vigorously during the titration. The end point of the titration is reached when the addition of a single drop produces a slight but definite colour change persisting for at least 15 seconds.

4.4 Acid Value :-

The number of mg of KOH required to neutralize the free fatty acids present in 1g of the sample.

$$\text{Acid Value} = \frac{5.61\ T}{W}$$

Where,

T = Volume of 0.1N alkali required for titration in ml and

W = weighting of sample taken.

4.5 Free Fatty Acids :-

The acidity of ghee is frequently expressed as the percentage of free fatty acids in the sample, calculated as oleic acid.

$$\text{Free fatty acids} = \frac{2.82T}{W}$$

4.6 Degree of Acidity :-

It is the total titratable acidity present in the sample expressed as percentage :

$$\text{Degree of acidity} = \frac{100N}{W}$$

Where,

N = the quantity of alkali used, expressed as ml of IN solution.

4.7 Accuracy of the Method :-

The maximum deviation between duplicate determination shall not exceed 0.2 degree of the acidity or equivalent.

5. Determination of soluble and insoluble volatile acids (Reichert or Reichert - Meissl, Polenske and Kirschner Values)

5.1 The method does not determine the total quantities of volatile fatty acids, soluble and insoluble in water, present in combination in fat. The amount of these acids actually determined by the process are dependent on strict adherence to the dimensions of the apparatus and the details of the procedure.

5.2 Definitions -

5.2.1 The Reichert - Meissl value (R.M. Value) is the number of ml of 0.1N aqueous alkali solution required to neutralize the water soluble steam volatile fatty acids distilled from 5g of ghee under the precise conditions specified in the method.

5.2.2 The Polenske value is the number of ml of 0.1N aqueous alkali solution required to neutralize the water insoluble steam volatile fatty acids distilled from 5g of ghee under the precise conditions specified in the method.

5.2.3 The Kirschner value is the number of ml of 0.1N aqueous alkali solution required to neutralize the water soluble steam volatile fatty acids which form water soluble silversalts distilled from 5g of ghee under the precise conditions specified in the method.

5.3 Apparatus :-

5.3.1 Graduated Cylinders - 100ml and 25ml capacities.

5.3.2 Pipette - 50ml.

5.3.3 The assembly of the apparatus for the distillation, details of the constituent parts are given below:

(a) Flat - bottom boiling flask (Polenske) - The flask shall be made of heat resistance glass and shall conform to the following details:

Volume contained to bottom of neck	310 ± 10ml.
Length of neck	75 ± 5mm
Internal diameter of neck	21 ± 1mm
Overall height	160 ± 5mm
Diameter of base	45 ± 5mm

(b) Still head - The still head shall be made of glass tubing of wall thickness 1.25 ± 0.25mm, and shall conform to the shape with the following dimensions:

A	180 ± 5mm
B	107.5 ± 2.5mm
C	80 ± 5mm
D	70 ± 5mm
E	20 ± 2mm
F	4 ± 1mm
G	(External diameter 37.5 ± 2.5mm of bulb)

Internal diameter of 80 $\pm$ 0.5mm

Tubing

Acute angle between 60 $\pm$ 2ºc

Stoping part of still-head

And vertical

A rubber stopper, fitted below the bulb of the longer arm of the still - head, and used for connecting it to the flask shall have its lower surface 10mm above the centre of the side hole of the still-head.

(c) Condenser - The condenser shall be made of glass and conform to the following dimensions -

Overall length	520 $\pm$ 5mm
Length of water jacket	300 $\pm$ 5mm
Length of widened	
Part above water jacket	70 $\pm$ 10mm
Wall thickness of widened part	1.25 $\pm$ 0.25mm
Internal diameter of widened part	20 $\pm$ 1mm
External diameter of inner tube within water jacket	12 $\pm$ 0.5mm
Wall thickness of inner tube	1.0 $\pm$ 0.2mm
Wall thickness of Outer Jacket	1.25 $\pm$ 0.25mm
External diameter of water jacket	30 $\pm$ 2mm

(d) Receiver - The receiver shall be a flask with two graduation marks on the neck, one at 100ml and the other at 110ml

(e) Asbestos board - An asbestos board of 120mm diameter and 6mm in thickness, with a circular hole of about 65mm in diameter shall be used to support the flask over the burner. During distillation the Polenske flask shall fit snugly into the hole in the board to prevent the flame from impinging on the surface of the flask above the hole. A new asbestos board may conveniently be prepared by bevelling the edge of the hole, soaking in water, moulding the edge with a flame, and drying.

(f) Gas burner - The burner should be sufficiently large to allow the distillation to be completed in the time specified.

5.3.3.1 The apparatus shall be supported on a retort stand.

5.3.4 Glass funnel - of approximate diameter 6mm.

5.4 Reagents -

5.4.1 Glycerol - 98% (w/w) conforming to AR grade of IS: 1796-1960.

5.4.2 Sodium Hydroxide -

50 per cent (w/w) solution. Sodium hydroxide is dissolved in an equal weight of water and the solution is stored in a bottle protected from carbon dioxide. The clear portion free from deposit is used.

5.4.3 Dilute Sulphuric Acid - approximately 25ml of concentrated sulphuric acid is diluted to 1:1 and adjusted until 40ml neutralize 2ml of the 50 per cent sodium hydroxide solution.

5.4.4 Glass Beads -approximately 1.5 to 2.0mm in diameter or ground pumice powder, passing through 250 micron IS sieve and remaining on 125 micron IS sieve.

5.4.5 Phenolphthalein Indicator - 0.5 per cent solution is 95 per cent (v/v) ethyl alcohol or rectified spirit.

5.4.6 Ethyl Alcohol - 95% per cent (v/v) neutralized to Phenolphthalein immediately before use or neutralized denatured spirit.

5.4.7 Sodium Hydroxide Solution - approximately 0.1N aqueous solution of sodium hydroxide of accurately determined strength.

5.4.8 Barium Hydroxide Solution - approximately 0.1N barium hydroxide solution of accurately determined strength (this solution is needed only if Kirschner value is to be determined).

5.4.9 Silver sulphate - powdered

5.4.10 Filter Paper - Whatman No.4 (or its equivalent) of 9cm diameter.

5.5 Procedure :-

5.5.1 Weigh 5.00 $\pm$ 0.01g of ghee into a Polenske flask. Add 20g of glycerol and 2ml of 50 per cent sodium hydroxide solution. Protect the burette containing the latter from carbon dioxide, and wipe its nozzle clean from carbonate deposit before withdrawing solution for the tests, reject the first few drops withdrawn from the burette. Heat the flask over a naked flame, with continuous mixing, until ghee, including any drops adhering to the upper parts of the flask, is saponified, and the liquid becomes perfectly clear, avoid overheating during this saponification. Cover the flask with a watch glass.

5.5.1.1 Make a blank test without ghee but using the same quantities of reagents and following the same procedure, again avoiding overheating during the heating with sodium hydroxide, such overheating would be indicated by darkening of the solution.

5.5.1.2 Measure 93ml of boiling distilled water, which has been vigorously boiled for 15 minutes, into a 100ml graduated cylinder. When the soap is sufficiently cool to permit addition of the water without loss, before the soap getting solidified, add the water, draining the cylinder for 5 seconds, and dissolve the soap. If the solution is not clear (indicating incomplete saponification) or is darker than light yellow (indicating overheating), repeat the saponification with a fresh sample of ghee.

5.5.1.3 Add two glass beads followed by 50ml of the dilute sulphuric acid, and connect the flask at once with the distilling apparatus. Heat the flask without boiling its contents, until the insoluble acids are completely melted, then increase the flame and distil 110ml in between 19 and 21 minutes. Keep the water flowing in the condenser at a sufficient speed to maintain the temperature of the issuing distillate to between 18 and 21ºC.

5.5.1.4 When the distillate reaches the 110ml mark, remove the flame and replace the 110ml flask by a cylinder of about 25ml capacity, to catch drainings. Close 110ml flask with its stopper and without mixing the contents, place it in water at 15ºC for 10 minutes so as to immerse the 110ml mark. Remove the flask from the water, dry from outside and invert the flask carefully avoiding wetting the stopper with insoluble acids. Mix the distillate by four or five double inversions, without violent shaking. Filter through a dry 9cm open texture filter paper (Whatman No.4) which fits snugly into the funnel. Reject the first runnings and collect 100ml in a dry volumetric flask, cork the flask and retain the filtrate for titration.

NOTE- The filtrate should be free from insoluble fatty acids. Where liquid insoluble fatty acids pass through the filter, receive the filtrate in a separating funnel and after separation, draw off the lower (aqueous) layer leaving behind insoluble acids which have risen to the surface. Add these to the main bulk of the insoluble acids.

5.5.1.5 Detach the still-head and wash the condenser with three successive 15ml portions of cold distilled water, passing each washing separately through the cylinder, the 110ml flask, the filter and the funnel, nearly filling the paper each time and draining each washing before filtering the next. Discard the washings, dissolve the insoluble acids by three similar washings of the condenser, the cylinder, and the filter, with 15ml of neutralized ethanol, collecting the solution in the 110ml flask and draining the ethanol after each washing. Cork the flask, and retain the solution for titration.

5.5.2 Reichert-Meissl or Soluble Volatile Acid Value - pour 100ml of the filtrate containing the soluble volatile acids into a titration flask, add 0.1ml of phenolphthalein indicator and titrate with the barium hydroxide solution until the liquid becomes pink, rinsing the 100ml flask with the nearly neutralized liquid towards the end of the titration (0.1N Sodium hydroxide solution may be used for the titration if the Kirschner value is not required).

5.5.2.1 If the kirschner value is to be obtained, the titration flask shall be dried before use; note the actual volume of barium hydroxide solution used, drain the 100ml flask into the titration flask, close with a cork and continue as in 5.5.4.

5.5.3 Polenske or Insoluble Volatile Acid Value - Titrate the alcoholic solution of the insoluble volatile acids after addition of 0.25ml of phenophthalein indicator with the 0.1N barium or sodium hydroxide solution until the solution becomes pink.

5.5.4 Kirschner value - Add 0.5g of finely powdered silver sulphate to the neutralized solutions reserved in 5.5.2.1. Allow the flask to stand in the dark for one hour with occasional shaking and filter the contents in the dark through a dry filter. Transfer 100ml of the filtrate to a dry Polenske flask, add 35ml of cold distilled water, recently boiled for 15 minutes, 10ml of the dilute sulphuric acid and 0.1g of pumice powder or two glass beads. Connect the flask with the standard apparatus and repeat the process as described above, that is, the distillation of 110ml in 19 to 21 minutes, the mixing (but without cooling for 10 minutes), and the filtration and the titration of 100ml of the filtrate with the barium hydroxide solution.

5.6 Calculations -

Reichert - Meissl value = 1.10 (T1 - T2)

Polenske value = T3 - T4

Kirschner value = $\frac{121\,(100 + T1)\,(T5 - T6)}{10{,}000}$

Where,

T1= Volume in ml of 0.1N barium or sodium hydroxide solution used for sample under 5.5.2,

T2= Volume in ml of 0.1N barium or sodium hydroxide solution used for blank under 5.5.2,

T3= Volume in ml 0.1N barium or sodium hydroxide solution used for sample under 5.5.3,

T4= Volume in ml of 0.1N barium or sodium hydroxide solution used for blank under 5.5.3

T5= Volume in ml of 0.1N barium hydroxide solution used for sample under 5.5.4, and

T6= Volume in ml of 0.1N barium hydroxide solution used for blank under 5.5.4.

Polenske values, and to a much slighter extent Reichert values, have been found to be low when determined at low barometric pressures, such as may occur at high altitudes. The following factors may be applied to values determined at a barometric pressure to convert them to the values determined at normal pressure.

Correct Reichert Value = $\frac{(\text{Observed Value} - 10)\ \text{Log}\ 760 + 10}{\text{Log p.}}$

$$\text{Corrected Polenske Value} = \text{Observed Value} \times \frac{760 - 45}{P - 45}$$

Where,

P = barometric pressure in mm of mercury at the place and time of determination.

5.7 Accuracy of the Method -

5.7.1 Reichert-Meissl Value - The maximum deviation between duplicate determination shall not exceed 0.5 units.

5.7.2 Polenske Value - The maximum deviation between duplicate determination shall not exceed 0.3 units.

5.7.3 Kirschner Value - The maximum deviations between duplicate determination shall not exceed 0.5 units.

6.0 Determination of Presence of Sesame Oil (Baudouin Test)

The development of a permanent pink colour with furfural solution in the presence of hydrochloric acid indicates the presence of sesame oil.

6.1 Reagents :-

6.1.1. Hydrochloric acid - fuming, sp. gr. 1.19.

6.1.2. Furfural solution -

2 per cent solution of furfural, which has been distilled earlier than 24 hours prior to the test, in rectified spirit.

6.2 Procedure :-

Take 5ml of the melted ghee in a 25ml measuring cylinder (or test tube) provided with a glass stopper and add 5ml of hydrochloric acid and 0.4ml of furfural solution. Insert the glass stopper and shake vigorously for two minutes. Allow the mixture to separate. The development of a pink or red colour in the acid layer indicates presence of sesame oil. Confirm by adding 5ml of water and shaking again. If the colour in acid layer persists, sesame oil is present. If the colour disappears, it is absent.

Chapter 38

Determination of Strength of Washing Solutions IS: 1479 (Part-V) 1962

1. **Introduction :-**

1.1 The strength of the detergent solution used depends on the use for which it is required. The strength of the washing solution is expressed in terms of sodium hydroxide (NaOH) but because of its caustic nature, sodium hydroxide is not often used direct especially at higher concentrations. An alkalinity equivalent to 0.5 per cent caustic soda (w/v) is generally used for washing bottles and milk cans. For dairy pipelines, valves and fittings, 1.5 per cent alkali (w/v) solution is used, while for cleaning heat exchangers when there is a likelihood of dried milk residues being left, 2.5 per cent solution (w/v) is used. The efficiency of detergent solutions is increased when used hot (50ºC to 70ºC).

1.2 The strength of the alkali in bottle washers has to be checked daily. One test shall be made at the beginning of the run after the alkali charge has been added and properly mixed. A second sample should be tested at the completion of the bottle washing operation. The alkali solution in can washer tank should be checked at the beginning and the end of the can-washing operation and as often in between as is necessary, to assure proper strength of the solution during the operation.

1.3 For determining the strength of detergent solutions either titration methods or measurement of pH changes may be employed for measuring pH the indicator methods are not so suitable due to the instability of these indicators at high pH (10 to 12). On the other hand, potentiometric method is quite suitable provided glass electrode is used. The titration method in various modifications is commonly used on account of its simplicity and low cost.

2. **Rapid Test (Qualitative):**

2.1 Reagents:-

2.1.1 Standard hydrochloric acid - 0.1N or 0.3N or 0.5N, depending upon the strength of detergent to be tested and accurately prepared.

2.1.2 Phenolphthalein indicator solution - Dissolve one gram of phenolphthalein in 100ml of 95 per cent ethyl alcohol by volume. Add 0.1N sodium hydroxide solution until one drop gives a faint pink colouration. Dilute with distilled water to 200ml.

2.1.3 Procedure :-Take 10ml of the detergent in an Erlenmeyer flask and add 12.5ml of 0.1N standard hydrochloric acids, if the detergent to be tested is equivalent to 0.5 per cent caustic soda. Add a few drops of the phenolphthalein indicator solution. If the mixture turns red, it indicates that the detergent solution has an alkalinity greater than 0.5 per cent caustic soda (w/v) and is of sufficient strength. If, on the other hand, the mixture remains colourless, it is an indication that the solution is too weak and that more detergent needs to be added. For examining 1.5 and 2.5 per cent detergent solutions, 0.3N and 0.5N hydrochloric acid shall be used and the above procedure followed.

3. **Rapid Test (Quantitative) :-** A convenient method for determining the strength of alkali solutions is to prepare a standard acid solution of such strength that the burette reading may be interpreted directly as percentage of alkali, when a particular quantity of sample is employed.

3.1 Reagents :-

3.1.1 Standard hydrochloric acid or standard sulphuric acid - 2.5N, accurately prepared.

3.1.2 Phenolphthalein indicator solution - prepare as given in 2.1.2.

3.2 Procedure :-

Measure exactly 10ml of the detergent solution to be tested in an Erlenmeyer flask, add 5 drops of phenolphthalein indicator solution. From a burette graduated in tenths, add the standard hydrochloric acid or sulphuric acid, drop by drop, until the solution becomes colourless. The number of millilitres used equals percentage of alkali, calculated as sodium hydroxide (NaoH).

4. **Laboratory Method :-** The method can be used to estimate either caustic alkali due to NaoH and part of carbonate and phosphate, or total alkali due to NaoH, all the carbonate and two-thirds of the phosphate by using two indicators.

4.1 Reagents :-

4.1.1 Standard sulphuric acid - 0.1N

4.1.2 Phenolphthalein indicator solution - prepare as given 2.1.2.

4.1.3 Methyl orange indicator solution - Dissolve 0.05g methyl orange in 100ml of water.

4.2 Procedure :-

Transfer 5ml of the test solution to a 250ml graduated flask and dilute to the mark with distilled water. Mix well, and transfer 50ml of the solution to an Erlenmeyer flask. Add a few drops of phenolphthalein indicator solution and titrate with the standard sulphuric acid to a colourless solution. Note the volume of acid used (A). To the same solution add now a few drops of methyl orange indicator solution and continue to titrate until a slight pink colour appears. Note the volume of acid used for this second titration (B) (as NaoH)

4.3 Calculation :-

Per cent by weight of caustic alkali due to NaoH and part

of carbonate and phosphate (expressed as NaoH) = A x 0.4

Per cent by weight of total alkali due to NaoH, all the carbonate and two-thirds of the phosphate = (A + B) x 0.4 (expressed as NaoH).

Chapter 39

Soft Water Analysis For Total Hardness

1. **Introduction:-** Total hardness in water is the sum of the concentrations of alkaline earth metal cations (eg. Ca++, Mg++). In all fresh waters nearly all the hardness is imparted by the calcium and magnesium ions which are in combination with bicarbonates and carbonates (temporary hardness) a part from sulphates, chlorides and nitrates. Determination of this parameter helps in deciding the use of water in dairy industry. Hard water can cause various problems in dairy e.g. scale formation, corrosion, increases the cost of steam generation, reduces the efficiency of cleaning etc. Therefore the hardness of water should be insured below 20 mg/litre. The easiest method for removing temporary hardness is boiling. For permanent hardness ion exchange is to be done i.e. replacement of calcium by sodium so that the insoluble form comes into soluble forms.
2. **Principle:-** Eriochrome black T forms wine-red complex compound with metal ions (ca++ and mg++). The di-sodium salt of EDTA extracts the metal ions from the dye-metal ion complex as colourless chelate complexes leaving a blue coloured aqueous solution of the dye.
3. **Requirements:-** Standard EDTA titrant (0.01M), Eriochrome black T indicator, ammonia buffer solution, titration assembly.
4. **Procedure:-** Take 50ml of the sample in a flask, add one ml of ammonia buffer and 5 drops of indicator solution. The colour of the sample turns wine-red. Titrate with EDTA solution, until a clear blue colour appears. Note the readings and calculate the total hardness.
5. **Calculation:-**

$$\text{Total hardness mgl-1 as CaCo3} = \frac{\text{ml of titrant used x 1000}}{\text{ml of sample}}$$

6. **Result:-** Express the total hardness in mgl-1 as CaCo3.

Chapter 40

Bacteriological Analysis of Water

1. General:-

The sanitary quality of water is of very great importance whether it is meant for drinking or for use in the food and dairy industries. If the water is polluted with faecal matter sewage or manure it is likely to contain many intestinal pathogens and communicate dangerous diseases like typhoid, cholera and dysentery to man. Natural waters (e.g. streams, wells, tanks) may also be contaminated with a variety of micro-organisms derived from the soil, vegetation and other sources. For assessing the sanitary quality of water it is generally examined for total bacterial content and for incidence of coliform bacteria, anerobic sporeformers or faecal streptococci. These three groups of bacteria are typical intestinal organisms and their presence in water serves as an index of pollution from sewage or faecal matter.

2. Sampling of water for bacteriological analysis :-

2.1 Great care is necessary in the collection of water samples for bacteriological analysis in order to secure truly representative samples from different sources of water and also to present any extraneous contamination of the samples at the time of the collection. The object of this excercise is to acquaint the students with the procedures to be adopted and the precautions to be taken in drawing samples of water from a municipal tap, stream and well for bacteriological analysis.

2.2 Material :-

2.2.1 Sterilized, ground glass stoppered bottle (8 oz capacity)

2.2.2. Methylated spirit or ethyl alcohol

2.3 Procedure :-

2.3.1 Taking samples from a tap

2.3.1.1 Allow the water to run for 3 or 4 minutes

2.3.1.2 Clean the inside and outside of the nozzle of the tap.

2.3.1.3 Sterilize the nozzle of the tap by heating it with a blow lamp or a piece of ignited cotton-wool soaked in methylated spirit or alcohol.

2.3.1.4 Again allow the water to run slowly for about a minute.

2.3.1.5 Hold the sample bottle in one hand near the tap, remove the stopper with the other hand, flame the mouth of the bottle quickly, bring the bottle below the running stream of water and when the bottle is nearly full take it out and quickly replace the stopper.

2.3.1.6 The above operations must be carried out quickly to prevent undue exposure of the bottle to atmospheric contamination and to avoid any water falling on the outside of the mouth dripping inside.

2.3.1.7 Transport the bottle immediately to the laboratory and keep it in a refrigerator till required for analysis.

2.3.2 Taking samples from a stream.

2.3.2.1 If the water is shallow, care should be taken not to disturb the soil at the bottom.

2.3.2.2 Hold the bottle in one hand and gently lower it (with the stopper on) into the stream as far below the surface as possible, remove the stopper with the other hand and as soon as the bottle is full replace the stopper when the bottle is still under water and then take it out of the stream.

2.3.2.3 Take care to ensure that no floating or suspended material falls into the bottle while taking the sample.

2.3.2.4 Wipe the exterior of the mouth and stopper with a clean cloth to prevent any water from dripping inside.

2.3.2.5 Transfer the sample to the laboratory at once for analysis.

2.3.3 Taking samples from a well.

2.3.3.1 Tie a strong string round the neck of a sample bottle and another around the stopper. Attach a weight to the bottle. Holding both the strings in the hand gently lower the bottle (with the stopper on) into the well to about a foot below the surface.

2.3.3.2 Take care to prevent the stopper coming out of the bottle while it is being lowered and to avoid any disturbance of the soil at the bottom while drawing the sample.

2.3.3.3 When the bottle is under water partially raise up the stopper by pulling the string attached to it so that water enters the bottle. When the bottle is filled put back the stopper into position by loosening the attached string and raise up the bottle.

2.3.3.4 Remove the strings attached to the bottle and stopper, wipe the exterior of the bottle with a clean cloth and transfer the sample to the laboratory for analysis.

3. Estimation of numbers of bacteria in water :-

3.1 General :- Information on the total numbers of bacteria in water as determined by the agar plate count technique, is helpful in estimating the general sanitary quality of the water supply and in the routine control of water purification processes. Samples containing less than 100 bacteria per ml are generally considered to be satisfactory and those having more than 1000 per ml as unsatisfactory. The relative proportion of faecal to other types of bacteria is however, of greater significance from the public health standpoint. The faecal types grow well at 37ºC in 24 to 48 hours while the other types, coming from plant sources, grow better at 20ºC - 22ºC but require a long incubation period of 3 to 5 days. The number of organisms from plant source should always be more in good quality water. The ratio of the total count at 37ºC to that obtained at 22ºC should be 1:10 and this ratio changes in case of contaminated water from faecal sources due to increase in the number of faecal organisms.

3.2 Material :-

3.2.1 Sample of water for analysis

3.2.2 Tryptone glucose agar tubes.

3.2.3 Dilution blanks (9ml saline or ringers solution)

3.2.4 Sterilized petridishes and 1ml bacteriological pipettes.

3.3 Procedure:-

3.3.1 Shake the water sample thoroughly by moving the bottle up and down 25 times.

3.3.2 Prepare 2 serial dilutions, 1 in 10 and 1 in 100 using the dilution blanks. Mix the dilution thoroughly by rotating the tubes between the palms.

3.3.3 Place twelve petridishes on the table, six for each series of one set at 37ºC and another at 22ºC.

3.3.4 Transfer 1 ml of the sample directly from the bottle (zero dilution), 1ml from dilution 1 (1/10) and 1ml from dilution 2 (1/100) into three separate petridishes (in duplicate). Mark the dilutions (0, 1 and 2) on the respective plates.

3.3.5 Melt twelve agar tubes, cool, pour into the plates and mix the agar with the inoculum by gently rotating the plates without allowing the agar to flow out.

3.3.6 When the agar has set invert the plates and incubate one series at 37ºC for 48 hours and the other at 22ºC for 72 hours.

3.3.7 At the end of the incubation period count the plates containing from 30 to 300 colonies. In the case of the zero dilution plates the number of colonies maybe counted even if it is less than 30.

3.3.8 Calculate the number of organisms per ml of water.

3.3.9 Record your observations and interpret the results

S. No.	Sample Source	Dilution	Bacterial Count atºC 37:22	Average bacterial count atºC 37:22	Total bacterial count atºC 37:22	Ratio of Count at ºC 37:22	Remarks
1							
2							
3							
4							

4 Presumptive coliform test for water and confirmatory tests for coliform's:-

4.A.1 General :- The presumptive coliform test is most commonly used for the detection of faecal contamination in water and evaluation of its sanitary quality. The organisms included in the group of coliform bacteria are all aerobic, gram negative, non-spore forming, rod-shaped bacteria which ferment lactose. With the formation of acid and gas in 48 hours at 37ºC when a sample of water is inoculated into bile salt lactose peptone broth and incubated at 37ºC. The production of acid and gas in 24/

48 hours is regarded as presumptive evidence of coliform contamination in water. A sample showing more than 2 coliform bacteria per 100ml of water is considered unsatisfactory. As the coliform group includes both faecal and non-faecal types of bacteria, it is necessary to conduct further confirmatory tests for the presence of faecal coli (E. coli) in water. The object of the exercise is to give the some practice in carrying out the presumptive coliform test for water by the serial dilution technique.

4.A.2. Material :-

4.A.2.1 Sample of tap water.

4.A.2.2 Mac Conkey's broth tubes (containing Andrade's) indicator and Duraham tubes, single as well as double strength.

4.A.2.3 Sterilized 10ml and 1ml pipettes.

4.A.2.4 9ml dilution blank.

4.A.3 Procedure:-

4.A.3.1 Transfer 1ml portions of the sample into 5 tubes of sinlge strength broth.

4.A.3.2 Prepare 1/10 dilution of the sample and transfer 1ml portions of the diluted sample into 5 tubes of single strength broth.

4.A.3.3 Transfer 10ml portions of the sample into 5 tubes containing 10ml of double strength medium.

4.A.3.4 There will thus be 3 sets of 5 tubes each containing 0.1ml, 1ml and 10ml, of the sample respectively.

4.A.3.5 Incubate the tubes at 37ºC for 24 hours and tubes showing no change should be incubated for another 24 hours.

4.A.3.6 Examine the tubes for production of acid (red colour) and gas in Durham tubes. Presence of acid and gas in three out of 5 tubes in any dilution is considered as a positive presumptive test for coliform bacteria. Note the highest dilution (containing the smallest amount of water) in which positive test is given .

4.A.3.7 Record your observations and interpret the results.

Sample No.	Source	Presumptive dilution broth tubes No. of positive 0.1ml. 1ml. 10 ml.	MPN/ 100ml. Remarks
1			
2			
3			
4			
5			
6			

4.B Confirmation test for coliforms:-

4.B.1 Select positive acid and gas tube from the previous experiments and subject them to confirm above tests. Two solid media are generally used in the test namely, eosin methylene blue (E.M.B.) agar medium and Endo agar medium. Typical colonies of coliform organisms will appear pink with dark centre and metallic sheen on E.M.B. agar. Endo agar produce coliforms colonies which appear red in colour and the growth will darken the medium to deep red.

4.B.2 Materials Required:-

4.B.2.1 E.M.B. agar or Endo agar.

4.B.2.2 Presumptive positive tubes of Mac Conkey's broth.

4.B.2.3 24 hours old nutrient broth culture of E.coli.

4.B.2.4 24 hours old nutrient broth culture of A. aerogenes.

4.B.3 Procedure:-

4.B.3.1 Pour 10 to 15ml of melted E.M.B. or Endo agar into the petridish and allow the media to set.

4.B.3.2 Make three sectors on lower dish by marking with glass marking pencil by inverting the petridish.

4.B.3.3. Move the inoculating needle slightly curved so that the streaking should ensure the presence of well isolated colonies.

4.B.3.4 Introduce the needle to the depth of ½ cm below the surface of the positive presumptive tubes.

4.B.3.5 Place the curved section of the needle on the agar surface in one segment and streak gently to avoid tearing of the medium.

4.B.3.6 Similarly streak a loopful of the culture of E.coli in second sector.

4.B.3.7 Streak the third sector with the culture of A. aerogenes.

4.B.3.8 Invert the plates and incubate at 37ºC for 24 hours.

4.B.3.9 Record your results.

Medium	Culture	Colour of medium	Colonies typical negative description
EMB Agar	From positive presumptive tube E.Coli A. aerogenes		
Endo Agar	Presumptive tube E.Coli A. aerogenes		

Chapter 41

Methods of Sampling and Test for Industrial Effluents

1. **General :-**

   This chapter covers the method of sampling and test for industrial effluents i.e. appearance, colour, odour, total suspended solids, pH value, temperature, dissolved oxygen, bio-chemical oxygen demand, oil and grease, chemical oxygen demand, total dissolved solid and chloride.

2. **Sampling :-**

   2.1 Point of Sampling - In those cases where the effluent at a specific point is to be tested, the question of choosing the point of sampling does not arise. However, where the composition of an effluent as finally discharged by a factory is to be ascertained, the point of sampling shall be the final outlet of the effluent treatment plant, where there is no treatment plant, it shall be the effluent outlet immediately outside the factory premises.

   2.2 Frequency of Sampling - When it is required to find out variations in the composition of an industrial effluents during a specified period, such as that of peak discharge, the samples shall be taken at short and appropriate intervals say every 5, 10, 15 or 30 minutes and analyzed. To study the average conditions over a cycle of operations or a period (usually 2 hours) or during the daily working period of the industry, the collection of composite sample shall be adopted. The composite sample shall be made by collecting at appropriate intervals samples from the common channel or drain at a point where the flow of the effluent is likely to be most representative of the entire volume and mixing.The volume of the individual samples shall be a fixed proportion of the volume of the effluent flowing at that time. The interval should depend upon the frequency of variation in the nature of the effluent and the volume of flow. Care shall be taken to take the samples in such a way as to maintain the true proportion of suspended solids. Samples shall not be taken by skimming the top or scraping the bottom. A point about one-third of the way from the bottom shall normally be selected. The samples shall be drawn gently without unnecessary aeration. In most cases, collection of samples every hour would be sufficient.

2.3 Sampling Instrument - Porcelain - lined or enamelled pails, in which the lining is unbroken or glass vessels shall be used for taking samples. The vessels used for taking the sample shall be wide-mouthed and small enough for the contents to be transferred quickly to the sample container without leaving behind any deposit or scum automatic sampling devices, if available, may be used.

2.3.1 Each individual sample shall be deposited in a receptacle of sufficient size to hold the entire composite sample. Clean and dry carboys, other large glass containers or enamelled buckets with lids without chipping may be used for pooling the sample.

2.4 Sample Containers:-

2.4.1 The quantity of sample required for analysis shall be taken from the composite sample after thorough mixing in order to keep the solids in suspension.

2.4.2 The sample for analysis shall be drawn in clean glass stoppered bottles, which shall be rinsed with a portion of the sample. New bottles shall be washed with acid and thoroughly rinsed with distilled water before being brought into use. About 2 to 3 litres of the sample will be required for analysis. The bottle containing the final sample shall be filled so that a small air bubble is present after closure to prevent leakage or even breakage arising from any subsequent changes in temperature. The stopper shall be firmly inserted and, if the sample is to be transported some distance, tied down to keep in position.

2.4.3 The label on the bottle shall bear the name of the sampling authority, details of the type of sample, place, date and time of sampling and, preferably, temperature and odour of sample.

2.5 Preservation of samples:-

2.5.1 The samples shall be kept at a low temperature (about 4ºC) during collection and thereafter.

2.5.2 No single method of preservation is applicable for the sample for all the tests. The tests for temperature, odour, dissolved sulphides and residual chlorine shall be carried out on the spot during sample collection. The test of the analysis shall also be carried out, preferably, immediately after collection. Storage at 3ºC to 4ºC in a well insulated ice box or refrigerator is the best way to preserve most samples till the next day. Where chemical preservatives are used as specified for individual tests, these shall be added to each portion of sample taken for the particular test and not to the entire sample.

2.5.3 Sample to be used for the estimation of total sulphide shall be preserved by adding 2ml of a 22 per cent solution of zinc acetate for litre of the sample.

2.5.4 Samples for the determination of phenol shall be acidified to a pH of less than 4-0 with orthophosphoric acid using methyl orange as indicator and preserved with 1g of copper sulphate per litre. The sample shall be kept around 5ºC to 10ºC and examined within 24 hours of collection.

2.5.5 Samples for the estimation of cyanides shall be

preserved by adding sodium hydroxide to raise the pH to 11.0 or above and stored in a cool place.

2.5.6 The addition of 5ml of concentrated nitric acid per litre of sample at the time of collection will help in the estimation of heavy metals by minimizing subsequent absorption of metals in the case of samples not already strongly acidic.

2.5.7 Samples for the estimation of oil and grease shall be preserved by adding 5ml of sulphuric acid (1:1) per litre of sample to inhibit bacterial activity.

3. General Precautions and directions for tests:-

3.1 Quality of Reagents - Unless specified other wise, pure chemicals and distilled water (See IS : 1070- 1960) shall be used in tests.

Note : ' Pure chemicals ' shall mean chemicals that do not contain impurities which affect the results of analysis.

3.2 It is important to obtain a representative sample. Appropriate methods of sampling are given in 2. However, in tests, where specific sampling procedures are prescribed, these shall be followed.

3.3 Many analytical procedures given are subject to interference is encountered or suspected and no specific procedure is laid down for overcoming it, steps shall be taken to eliminate the interference without adversely affecting the analysis itself.

3.4 Where the range of amount of the substance being determined is stated, it is usually given in either micrograms (µg) or milligrams. The procedures laid down for the various tests cover the stated ranges, but may, if necessary, be modified either by taking a smaller volume of the sample and diluting with distilled water or by taking a larger volume of the sample and evaporating it.

3.5 For determinations for which calibrated glass discs are available, these may be used for routine examination provided instructions of the manufacturer are followed. But in case of dispute, test methods are prescribed in this standard shall be followed.

4. **Appearance:-**

4.1 Observe and record the general appearance of the sample, the tendency to foam, and presence of visible oil, tar or other floating, suspended or settled matter.

5. **Colour:-**

5.1 Observe and record the colour of the sample contained in the receptacle.

6. **Odour:-**

6.1 Gently shake the sample, smell, and record the true odour of the sample at the mouth of the bottle as musty, septic, aromatic, chlorinous, disagreeable etc.

7. **Total suspended solids:-**

7.0 Outline of the Method - Suspended matter is determined by filtering the sample through an asbestos pad in a Gooch crucible.

7.1 Reagent:-

7.1.1 Asbestos Cream - Make a cream of acid - washed medium - fibre Gooch asbestos with water. Add one litre of water for every 15g of asbestos. If the asbestos contains too much fine powder, remove the latter by repeated decantation.

7.2 Procedure :

7.2.1 Make carefully an asbestos mat in the Gooch crucible by adding sufficient asbestos cream to produce a mat about 3mm thick. In preparing the mat, first fill the crucible with well-mixed asbestos cream, let stand for about two minutes to allow the heavier particles to settle down and then apply suction to the same extent as will be used for filtering the sample. Wash the mat with water with the suction on by filling and drawing through. Dry the crucible with the asbestos mat in an oven at 103ºC to 105ºC for one hour, cool in a desiccator and weigh.

7.2.2 Filter the sample through the weighed Gooch crucible after moistening with a few drops of water. Add successive increments of 10ml of the well-shaken sample for filtration using suction. Add each increment of sample before the mat becomes dry. The use of a pipette with an orifice wide enough to prevent clogging with suspended matter is recommended. Continue successive 10 ml additions of the sample until the filtration becomes inconveniently slow or until about 10 to 20mg of suspended matter has been filtered. Carefully wash the mat with 10ml of water to remove soluble salts. Continue suction until draining is complete. Dry the crucible in an oven at 103ºC to 105ºC for one hour, cool to room temperature in a desiccator and weigh.

7.3 **Calculation :**

$$\text{Total suspended solids, mg/l} = \frac{1000W}{V}$$

Where,

W= weight in mg of the suspended matter,

V= volume in ml of the sample taken for filtration.

7.3.1 Express the result to the nearest 5 mg/l.

8. **Particle size of suspended solids:-**

8.0 **General**

8.0.1 Outline of the Method :- The particle size of suspended solids is determined by wet screening of the freshly drawn sample through the specified sieve.

8.0.2 Since the suspended solids in the sample are likely to coalesce on keeping, the test should be carried out on the spot, leaves, twigs and other wind - blown debris, which are extraneous to the sample, should be removed.

8.1 **Apparatus:-**

8.1.1 Sieve - 850 – micron IS sieve or any other IS sieve

if so specified in the relevant standard.

8.1.2 Enamelled pail of diameter slightly bigger than that of the sieve.

8.2 **Procedure** :- Hold the sieve in one hand and, with the other, pour gently on the mesh surface of the sieve one litre of the well-mixed sample. Pour the sample in such a way that it covers the entire mesh surface. If necessary, create a vibration while sieving the sample by a gentle rocking motion of the hand holding the sieve. Fill the enamelled pail with fresh water. Then, holding the screen on opposite sides with the two hands bring it to the surface of water in the enamelled pail and wet screen by jigging (up and down motion). Take care to see that, while jigging, the sieve is dipped in the water only to half its depth and there is no overflow from the mesh through the sides as suspended solids would be washed out without passing through the screen. If necessary wash the material on the screen with a fine jet of water from a wash-bottle until all lodged particles are loosened.

8.2.1 The suspended matter shall be considered passing through the sieve only if no residue is left on it.

9. pH VALUE

9.0 **General:** The pH value may be determined either electrometrically as given in 9.1 or colorimetrically as given in 9.2 . The electrometric method is more accurate but requires special apparatus . The colorimetric method is simple and requires less expensive apparatus, and is sufficiently accurate for general work. It is, however subject to interference by colour, turbidity, high saline content, free chlorine and various oxidants and reductants. In case of dispute the electrometric method shall be considered as the referee method.

9.1 Electrometric Method - The determination may be made with any pH meter provided with a glass electrode, using instructions of the manufacturer. Express the result to the nearest 0.1 unit.

9.2 Colorimetric Method.

9.2.1 Reagents - A series of indicators and buffer solutions are required for this method. The methods of preparation of indicators are given in Appendix A. The buffer solutions shall be prepared as prescribed in IS : 3225 - 1965.

9.2.2 Procedure - Take 100ml of the sample in a hard glass tube and determine the approximate pH by using a universal indicator. Repeat using a solution of the indicator (about 1/20 of the volume of the liquid being tested) which corresponds to the approximate pH found above, compare the colour produced with a series of buffer solutions of known pH, each containing the same proportion of the indicator. Report as pH, the pH of that buffer solution which matches with that of the sample, to the nearest 0.1 unit.

10 Temperature:-

10.1 Use a good grade mercury filled centigrade thermometer. Measure the temperature at the time and place where the sample is collected by dipping the thermometer into the sample until the whole bulb is immersed. Record the temperature to the nearest 0.5 degree centigrade.

10.1.1 During the test, measure the atmospheric temperature at the site and record.

11. Dissolved oxygen:-

11.0 Outline of the Method - All the methods for the determination of dissolved oxygen are based on the original winkler procedure. The principle is that of precipitation of mangnous hydroxide in a bottle filled with the sample brought about by the addition of a solution of manganous sulphate followed by alkaline potassium iodide solution. The oxygen present in the sample quickly reacts with the manganous hydroxide forming a brown precipitate of higher hydroxides. On subsequent acidification, iodine in an amount equivalent to the oxygen contained in the sample is liberated. The quantity of iodine liberated is estimated by titration against a standard solution of sodium thiosulphate, using starch as indicator.

11.0.1 The winkler method is subject to interference due to various ions and compounds contained in the sample. Suitable modifications of the method are adopted to correct for these interferences. The choice of the exact modified procedure will depend upon the nature of the sample and the interference present. The application of these procedures and the methods are given below :

Sl. No.	Procedure	Applicability
(i)	Winkler method	When no interfering substances are present.
(ii)	Alsterberg (sodium azide) modification	When not more than 0.1 mg/l of nitrite nitrogen and not more than 1 mg/l of ferrous iron are present and in the absence of other reducing or oxidizing agents. In the presence of 5mg/l or more of ferric iron, potassium fluoride is added. When potassium fluoride is added, the method is applicable in the presence of 100 to 200 mg/l of ferric iron.
(iii)	Rideal-Stewart (permanganate) modification	In the presence of ferrous iron only. If ferrous or ferric iron is more than 10 mg/l, potassium fluoride is added before acidifying.
(iv)	Alkali hypochlorite modification	In the presence of sulphite, thiosulphate, polythionate, free chlorine and hypo chlorite. However, the results obtained with this method cannot be relied upon to give accurate results in the determination of dissolved oxygen as this procedure gives somewhat low results.
(v)	Short-Theriault modification	In the presence of organic matter easily oxidized at the pH of the alkaline iodide treatment.
(vi)	Alum flocculation modification	In the presence of high amounts of suspended solids.

11.1 Winkler Method:-

11.1.1 Reagents:

11.1.1.1 Manganous sulphate solution - Dissolve 480g of manganous sulphate tetra hydrate ($MnSO4 . 4H_2O$) in water, filter if not clear, and make up to one litre. The solution should not liberate more than a trace of iodine when added to an acidified potassium iodide solution.

11.1.1.2 Alkaline iodide solution - Dissolve 500g of sodium hydroxide (or 700g of potassium hydroxide) in its own weight of water and allow to cool. The caustic solution should be virtually free from carbonate. Dissolve separately 150g of potassium iodide in a small quantity of freshly boiled and cooled water and add this solution to the caustic solution. Dilute the mixture to one litre.

11.1.1.3 Concentrated sulphuric acid

11.1.1.4 Standard sodium thiosulphate working solution - exactly 0.25N, freshly standardized against potassium dichromate. One millilitre of this solution is equivalent to 0.2mg of oxygen.

11.1.1.5 Starch indicator solution - Triturate 5g of starch and 0.01g of mercuric iodide with 30ml of cold water and slowly pour it with stirring into 1 litre of boiling water. Boil for 3 minutes. Allow the solution to cool and decant off the supernatant clear liquid.

11.1.2 Procedure - Remove the stopper from a 250 to 300ml bottle containing the sample and add 1.5ml of manganous sulphate solution followed by 1.5ml of alkaline iodide solution, keeping the tip of the pipette in each case well below the surface of the liquid. Carefully replace the stopper without the inclusion of air bubbles and thoroughly mix the contents by inverting and rotating the bottle several times, allow the precipitate formed to settle. When the precipitate settles leaving a clear supernatant above the manganese hydroxide floc, repeat mixing a second time and allow to settle. When further settling produces atleast 100ml of clear supernatant, carefully remove the stopper and immediately add 2ml of concentrated sulphuric acid by running the acid down the neck of the bottle, restopper and mix well to ensure uniform distribution of iodine in the bottle. Take 200ml of the solution and titrate immediately against standard sodium thiosulphate solution, adding 1ml of starch indicator solution when the colour becomes pale yellow and completing the titration to the disappearance of the blue colour. The error due to the displacement of sample by reagents is insignificant.

11.1.3 Calculation

Dissolved oxygen, mg /l = V

Where,

V= volume in ml of standard sodium thiosulphate solution used in the titration.

11.2 Alsterberg (Azide) Modification :

11.2.1 Reagents - Use all the reagents given in 11.1.1 except alkaline iodide solution, which is to be replaced by alkaline iodide-sodium azide solution and potassium fluoride reagent.

11.2.1.1 Alkaline iodide-sodium azide solution - Dissolve 10g of sodium azide in 40ml of water. Add this with constant stirring to the cool alkaline iodide solution prepared as in 11.1.1.2 but made upto 950ml.

11.2.1.3 Potassium fluoride solution - Dissolve 40g of potassium fluoride (KF. $2H_2O$) in 100ml of water.

11.2.2 Procedure - The procedure given in 11.1.2 shall be followed but, instead of alkaline iodide solution, alkaline iodide-sodium azide solution shall be used. When 5 mg/l of ferric iron is present, add 1ml of potassium fluoride solution before acidifying the sample and titrate immediately after acid addition and mixing.

11.2.3 Calculation - Calculate as in 11.1.3

11.3 Rideal Stewart (Permanganate) Modification :

11.3.1 Reagents - Use all the reagents given in 11.1.1, potassium fluoride solution given in 11.2.1.2 and, in addition, the following reagents.

11.3.1.1 Potassium permanganate solution - Dissolve 6.3g of potassium permanganate in water and make up to 1 litre.

11.3.1.2 Potassium oxalate solution - Dissolve 2g of potassium oxalate ($K_2C_2O4\text{-}H_2O$) in 100 ml of water.

11.3.2 Procedure - Remove the stopper of the bottle containing the sample, add below the surface, with a 1ml graduated pipette, 0.70ml of concentrated sulphuric acid followed by sufficient potassium permanganate solution (about 1ml) to produce a violet tinge which persists for 5 minutes. Avoid large excess of permanganate, stopper and mix by inversion. After 5 minutes remove the excess permanganate by adding, 0.5ml portions of potassium oxalate solution and mixing. Allow 5 minutes intervals after each addition of oxalate. Then proceed as in 11.1.2, using 3ml of alkaline iodide solution instead of 1.5ml. When either ferrous or ferric iron is present in excess of 10 mg/l, add 1ml of potassium fluoride solution immediately after permanganate addition. The titration after final acidification should be carried out without delay.

11.3.3 Calculation - Calculate as in 11.1.3

11.4 Alkali Hypochlorite Modification:

11.4.1 Reagents - Use all the reagents given in 11.1.2, substituting alkaline iodide solution with alkaline iodide-sodium azide solution (see 11.2.1.1). In addition, use the following reagents.

11.4.1.1 Alkali hypochlorite solution (2N) - prepared by passing chlorine gas through a 2.1 N sodium hydroxide solution, with cooling. One millilitre of this solution will require, on acidification in the presence of potassium iodide, about 20ml of 0.1 N sodium thiosulphate solution for titration. Check the strength of the solution every week.

11.4.1.2 Potassium iodide solution (IN) - Dissolve 17g of potassium iodide in water and make up to 100ml. Preserve by adding 1ml of 1N sodium hydroxide solution.

11.4.1.3 Sulphuric acid - 1:9 (v/v).

11.4.1.4 Sodium sulphite solution (0.1N) - Dissolve 6.3g of anhydrous sodium sulphite or 12.6g of sodium sulphite heptahydrate in water and make up to 1 litre. Do not use the solution when its strength goes down to less than 80 percent of the original.

11.4.1.5 Potassium biniodate solution (0.1N) - Dissolve 3.429g of potassium biniodate (KH(IO3)2) in water and make up to 1 litre in a volumetric flask.

11.4.2 Procedure - Remove the stopper of the bottle containing the sample and add 0.2ml or just sufficient quantity of alkali hypochlorite solution to oxidize the sulphite. Stopper and mix by inversion for 20 to 30 seconds. Add 1ml of potassium iodide solution and acidify with 1ml or more of sulphuric acid. Mix by inversion. Add

0.2ml of starch indicator solution and destroy the iodide liberated with sodium sulphite solution. Restore the blue tinge with 0.1ml portions of 0.1N potassium biniodate solution. Proceed further as in 11.2.2 with the difference that 3ml of alkali iodide sodium azide solution has to be added instead of 1.5ml.

11.4.3 Calculation - Calculate as in 11.1.3

11.5 Short-Theriault Modification :-

11.5.1 Reagents - Use all the reagents given in 11.1.1, substituting alkaline iodide solution with alkaline iodide-sodium azide solution (see 11.2.1.1).

11.5.2 Procedure - Remove the stopper of the bottle containing the sample, add 2ml of manganous sulphate solution followed by 2ml of alkali iodide-sodium azide solution, stopper and mix by inversion for 20 seconds. Immediately add 2ml of concentrated sulphuric acid before the precipitate settles, remix and titrate as prescribed in 11.1.2.

11.5.3 Calculation - Calculate as in 11.1.3.

11.6 Alum Flocculation Modification :-

11.6.1 Reagents - Use all the reagents given in 11.2.1 and, in addition the following reagents.

11.6.1.1 Alum solution - Dissolve 10g of potassium aluminium sulphate in water and dilute to 100 ml.

11.6.1.2 Ammonium hydroxide - concentrated.

11.6.2 Procedure - Collect a sample in a glass stoppered bottle of 500 to 1000ml capacity, using the same precautions as are necessary for the 300ml sample. Add 10ml of alum solution followed by 1 to 2ml of ammonium hydroxide. Stopper and invert gently for about one minute. Allow to settle about 10 minutes and then siphon the clear supernatant into a 250 tó 300ml dissolved oxygen bottle until it overflows. Avoid aeration and keep the siphon submerged atleast 20 cm. Then follow the procedure and calculate the result as given in 11.2.11.3, 11.4 or 11.5, as appropriate.

12. Bio-chemical oxygen demand:-

12.0 Out line of the Method - Biochemical oxygen demand (BOD) is the quantity of oxygen required by a definite volume of the liquid effluent for oxidizing the organic matter contained in it by micro-organisms under specified conditions for its determination, the dissolved oxygen content of the sample, with or without dilution, is measured before and after incubation at 20ºC for 5 days.

12.1 Apparatus:-

12.1.1 Glass stoppered Bottles - Narrow-neck bottles of about 250ml capacity, with suitable water sealing.

12.2 Reagents

12.2.1 Sodium Hydroxide solution – approximately 1N.

12.2.2 Hydrochloric Acid - approximately 1N.

12.2.3 Sodium Sulphite solution - Dissolve 1.5g of anhydrous sodium sulphite in 1 litre of water. Prepare fresh solution daily for use.

12.2.4 Dilution water - distilled water of good quality free from metals, particularly copper, and aerated.

12.2.5 Phosphate buffer solution - Dissolve 8.5g of potassium dihydrogen phosphate (KH_2 PO4), 21.75g of dipotassium hydrogen phosphate (K_2HPO4) 34.4 g of disodium hydrogen phosphate (Na_2 HPO4. $7H_2O$) and 1.7g of ammonium chloride in about 500ml of water and dilute to 1 litre. The pH of this solution should be 7.2.

12.2.6 Magnesium Sulphate Solution - Dissolve 22.5g of magnesium sulphate ($MgSO_4$. $7H_2O$) in water and dilute to 1 litre.

12.2.7 Calcium Chloride solution - Dissolve 27.5g of anhydrous calcium chloride in water and dilute to 1 litre.

12.2.8 Ferric Chloride solutions- Dissolve 0.25g of ferric chloride ($FeCl3. 6H_2O$) in water and dilute to 1 litre.

12.2.9 Seeding Material - Supernatant liquor of domestic sewage stored for 24 to 36 hours at 20ºC. In the case of industrial effluents containing organic compounds which are not easily oxidized by sewage seed, the receiving water collected about 3.5 km below the discharge point may be used.

12.3 Procedure :-

12.3.1 Samples containing acidity or caustic alkalinity should be neutralized to pH about 7.0 with sodium hydroxide solution or hydrochloric acid respectively by adding a predetermined quantity.

12.3.2 Samples containing residual chlorine or chloramines should be dechlorinated if chlorine is not dissipated on standing for 2 hours. To dechlorinate, first determine the quantity of sodium sulphite solution required for a known aliquot of the sample by titration to starch-iodide end point after acidifying the sample with acetic acid (1:1) or sulphuric acid (1:50) followed by 10ml of 10 per cent potassium iodide solution. Then add to the requisite volume of the sample the predetermined quantity of sodium sulphite, avoiding any excess, and check for the absence of chlorine after 20 minutes.

12.3.3 Samples containing toxic substances in large amounts would require special treatment. However, the effect of small amounts may be overcome by using the proper dilution so that toxicity is removed and the maximum BOD value is obtained. If increasing dilutions show increasing BOD, the dilution should be increased to a level where BOD levels off at a maximum.

12.3.4 To check the quality of the dilution water and the effectiveness of the seed, determine the BOD of a standard solution of 300 mg/l of either glucose or glutamic acid in the dilution water. Standard glucose solution should show a BOD of 224 $\pm$ 10 mg/l and glutamic acid 217 $\pm$ 10 mg /l.

12.3.5 Store the dilution water at 20ºC and use when near that temperature. Take the desired volume of dilution water required for the test sample and add, for every 1 litre of water, 1ml each of phosphate buffer solution, magnesium sulphate solution, calcium chloride solution and ferric chloride solution. Seed the dilution with seeding material. The quantity of seeding material (0.1 to 1 per cent of settled sewage or 1 to 5 per cent receiving water) added should be such that oxygen depletion in the dilution water control is between 0.2 and 0.8 mg/l after incubation at 20ºC for 5 days.

12.3.6 Prepare as follows several dilutions of the sample (usually 0.1 to 1.0 per cent for strong industrial effluents and 5 to 25 per cent for treated effluents) so as to obtain a depletion of at least 2 mg/l of dissolved oxygen after incubation for 5 days. In the case of dilutions greater than 1:100, prepare a 10 per cent primary dilution in a volumetric flask and from this make the final dilutions.

12.3.7 Siphon carefully the prepared seeded dilution water into a graduated 1000ml measuring cylinder and fill to the 500ml mark. Add the requisite quantity of the carefully well mixed sample to make the particular dilution and fill with dilution water to 1 litre. Mix thoroughly but gently with a plunger type of rod without entraining air. Siphon the dilution in to two glass-stoppered bottles, fill completely and stopper. Prepare succeeding dilutions of lower concentrations in the same manner. Determine the initial dissolved oxygen concentration in one of the two bottles of each dilution by the appropriate method given in 11. water seal the other bottles and incubate at 20ºC for 5 days. At the same time, siphon the dilution water alone into two glass-stoppered bottles and determine the blank in one and incubate the other at 20ºC for 5 days. After incubation for 5 days, determine the dissolved oxygen in the dilutions and the blank in the same manner as the initial dissolved oxygen content.

12.4 Calculation :-

Biochemical oxygen demand (5 days at 20ºC), mg/l $= \dfrac{(D1-D2) - (C1-C2)}{P}$

Where

D1 = initial dissolved oxygen content of the diluted sample,

D2 = Dissolved oxygen content of the diluted sample after incubation,

C1 = initial dissolved oxygen content of the seeded dilution water,

C2 = dissolved oxygen content of the seeded dilution water after incubation,

F = ratio of the seed in the sample to that in the control, that is, per cent seed in D1 divided by percent seed in C1, and

P = decimal fraction of the sample used.

12.4.1 Express the result to the nearest 0.2 mg/l.

13. Oils And Grease:-

13.0 Outline of the method – The oils and grease are extracted by an organic solvent. The solvent is distilled off and the weight of the extracted matter determined.

13.1 Sampling - The most satisfactory method of sampling two-phase liquids is to use a sampling tube that is capable of withdrawing a complete section of the effluent as it flows in a rectangular culvert or trough, in most instances, however, the effluent will have to be sampled from the outfall of a pipe or from a stream and, in these circumstances, some of it should first be collected in a large cylindrical vessel having a capacity of 10 to 15 litres. A sectional sampling tube should be used to withdraw the test sample from this. A sampling tube, suitable for sampling effluents that do not contain highly viscous matter (for example, tar). The sampler consists of a heavy-gauge brass tube, 1 meter long, with an outside diameter of 40mm over one end of the tube is filled a brass bucket made from a piece of tube, 50mm long and scaled at one end. The bucket has an internal diameter 1.5mm greater than the outside diameter of the main tube. To opposite sides of the bucket are brazed two brass-rods, 6mm in diameter, which pass through guides brazed to the sides of the main tube. The rods are so arranged that the top of the bucket can be withdrawn to a distance of not less than 10cm from the bottom of the main tube and they guide the bucket into a position covering the end of the tube when it is pushed back again. A suitable spring catch is provided on one of the guide rods so that the bucket is automatically locked into the top position when it is raised to its highest point. The open end of the sampling tube is fitted with a rubber bung.

13.1.1 To take a sectional sample, the spring catch is released and the bucket is drawn as far as possible away from the end of the main tube. The rubber bung is withdrawn from the other end. The tube is lowered vertically through the liquid to be sampled until the bottom of the bucket rests on the bottom of the culvert or of the vessel that has been filled with the effluent. The main tube is then pushed down, guided by the brass rods, to the limit of its travel, whereupon the spring catch locks the bucket in the raised position covering the end of the tube. The rubber bung is tightly inserted in the open end and the tube is withdrawn. The outside of the sampler is wiped free from the adhering liquid, the bucket and the lower part of the tube are inserted into a wide-mouthed bottle of suitable capacity and the rubber bung is removed. The sample section of the liquid will blow in to the bottle, leaving a small quantity of liquid in the bucked. The tube is then tilted, so that this liquid is added to the main bulk of the sample. The operation is repeated until a sufficient quantity has been collected. Atleast 25 mm of air space should be left between the top level of the liquid and the stopper of the bottle.

13.2 Apparatus:-

13.2.1 Separating Funnel - 500ml capacity. The stopper or stop-cock should not be lubricated with matter soluble in petroleum ether.

13.3 Reagents:-

13.3.1 Magnesium sulphate solution - Dissolve 1 g of magnesium sulphate heptahydrate in 100ml of water.

13.3.2 Milk of lime - Mix 2g of calcium oxide with water into a paste and dilute the suspension to 100ml.

13.3.3 Light petroleum (petroleum ether) - boiling range 40ºC to 60ºC.

13.3.4 Dilute Hydrochloric Acid - 1:3.

13.3.5 Sodium-sulphate-anhydrous.

13.4 Procedure :-

13.4.1 Take 250ml, or an aliquot containing 50 to 150mg, of extractable matter of the well-mixed sample in a beaker. If a noticeable layer of floating matter is present in it, carefully transfer as much of it as possible by decantation into a separating funnel. Draw in to the beaker containing the residual portion of the sample any liquid that separates out in the funnel. To the sample in the beaker, add 5ml of magnesium sulphate solution. Stir in a rotatory direction with a glass rod and add continuously small amounts of milk of lime until flocculation occurs (see note). Continue stirring for 2 minutes, withdraw the glass rod and wash it down in the separating funnel with a small quantity of light petroleum. Allow the precipitate in the beaker to settle for 5 minutes. When it has settled completely, siphon off the clear supernatant liquid to within about 1cm of the top of the sediment. Allow any remaining floating oil to be in the beaker itself. Dissolve the precipitate in the beaker with dilute hydrochloric acid and transfer the contents to the separating funnel, taking care not to transfer any large adventitious solids like twigs, leaves, etc. Rinse the beaker with about 50ml of light petroleum and add this to the liquid in the funnel. Shake the funnel continuously, but not vigorously, for one minute. Allow the liquid layers to separate. Draw the aqueous layer into another separating funnel and extract again with a fresh 50 ml portion of light petroleum. Reject the aqueous layer and combine the petroleum extracts.

Note - Some effluents do not readily flocculate with lime, in such cases, determine the suitable flocculating agents by a preliminary trial and add them. The following flocculating agents are suggested :

a) Aluminium sulphate - one percent solution with pH adjustment of the sample.

b) Ferric chloride - one percent solution and ammonium hydroxide.

c) Zinc acetate - 10 per cent solution and sodium carbonate 5 per cent solution.

13.4.2 Add to combined petroleum extracts 2g of powdered anhydrous sodium sulphate and shake intermittently over a period of about 30 minutes. Filter through a small size filter paper (Whatman No. 30 or equivalent) collecting the filtrate in a dry, weighed wide-neck flask of 250ml capacity. Wash the paper with two successive 20ml portions of light petroleum and collect the filtrate in the flask. Distil off most of the light petroleum from the filtrate in the flask and finally, evaporate the last traces in a current of warm air. Keep on a waterbath for 10 minutes, wipe the outside dry with a filter paper, cool in a desiccator and weigh. The difference in weight is the weight of the residue. (If, after the solvent has evaporated, the residue contains water, add 2ml of acetone and evaporate on a waterbath. Repeat the acetone addition and evaporation until the contents are free of water.)

13.5 Calculation:-

$$\text{Oil and grease, mg/l} = \frac{1000W}{V}$$

Where

W= weight in mg of the residue, and

V= volume in ml of the sample taken for the test.

13.6 Express the result to the nearest milligram.

14. Chemical Oxygen Demand (COD):-

14.0 Outline of the Method - This is determined by refluxing the sample with an excess of potassium dichromate in acid conditions and estimating by titration the amount of dichromate consumed.

14.1 Interference - Unstable samples should be tested without delay and samples containing settleable solids should be homogenized by suitable means for ease of representative sampling. Initial dilutions in volumetric flasks should be made on those samples having a high COD, in order to reduce the error which is inherent in measuring small sample volumes. Chlorides are quantitatively oxidised by this procedure when silver sulphate is not used as a catalyst. In this case, a correction should be applied by determining chlorides on a separate sample and subtracting the calculated oxygen demand of the chlorides from the result. Sincc 1 mg/l of chloride will consume 0.23 mg/l of oxygen, the correction is mg/l of chloride x 0.23.

14.2 Reagents:-

14.2.1 Standard potassium Dichromate solution - 0.25 N. Dissolve 12.259g $K_2Cr_2O_7$, Primary standard grade, previously dried at 103ºC for 2 hours, in distilled water and dilute to 1000ml.

14.2.2 Concentrated sulphuric acid- See is : 266-1961 specification for sulphuric acid (revised). Containing 22g Silver Sulphate (Ag_2SO4) per 9-lb bottle.

14.2.3 Standard Ferrous Ammonium Sulphate Solution- Dissolve 97.5g Fe $(NH_4)2(SO_4)2.6H_2O$ in distilled water. Add 20ml conc.H_2SO_4 cool and dilute to 1,000ml. This solution must be standardized against the standard potassium dichromate solution daily.

14.2.3.1 Standardization - Dilute 10.0ml standard potassium dichromate solution to about 100ml. Add 30 ml conc H_2SO_4 and allow to cool. Titrate with the ferrous ammonium sulfate titrant, using

2 or 3 drops (0.10 - 0.15ml) ferroin indicator.

$$\text{Normality} = \frac{\text{ml } K_2Cr_2O_7 \times 0.25}{\text{ml Fe } (NH_4)2\ (SO_4)2}$$

14.2.4 Ferroin indicator solution - Dissolve 1.485g of 1,10 phenanthroline (monohydrate), together with 0.695g of ferrous sulphate ($FeSO_4.7H_2O$) in distilled water and dilute to 100ml.

14.2.5 Silver sulphate

14.3 Procedure:-

14.3.1 Place a 50ml sample, or an aliquot diluted to 50ml with distilled water, in a 300ml round bottom flask fitted with ground-glass joint for attaching a condenser, and add 25ml of standard potassium dichromate solution. Carefully add 75 ml of concentrated sulphuric acid, mixing after each addition.

Caution - The mixture shall be thoroughly mixed before heat is applied. If this is not done, local heating occurs in the bottom of the flask and the mixture may be blown out.

14.3.2 Attach the flask to the condenser and reflex the mixture for 2 hours. Pumice granules or glass beads should be added to the reflux mixture to prevent bumping. Cool and then wash down the condenser with about 25ml of distilled water. In many cases, the 2-hour reflux period is not necessary. Therefore, with particular samples, the reflux period necessary to give the maximum COD should be determined and the shorter period of refluxing may be permissible.

14.3.3 Transfer the contents to a 500ml conical flask, washing out the reflux flask 4 to 5 times with distilled water. Dilute the mixture to about 350 ml and titrate the excess potassium dichromate with standard ferrous ammonium sulphate solution, using ferroin indicator. Generally 2 to 3 drops of the indicator are used. The colour change is sharp, changing from a blue-green to a reddish-blue. The end point, however, will not be as sharp as in the standardization of the reagents because of the lower acid concentration. For this reason, it is necessary that the sample be diluted to at least 350ml before the titration is carried out. A blank consisting of 50ml of distilled water instead of the sample, together with the reagents, is refluxed in the same manner.

Note - More complete oxidation of many organic compounds, such as straight-chain alcohols and acids, may be obtained by the use of silver sulphate as a catalyst. One gram of silver sulphate is added directly to the mixture before refluxing.

14.4 Calculation :-

$$\text{Chemical oxygen demand, mg/l} = \frac{(A\text{-}B)\ N \times 8000}{V}$$

Where

A = Volume in ml of ferrous ammonium sulphate solution used in the titration in the blank,

B= Volume in ml of ferrous ammonium sulphate solution used in the titration with the sample,

N= Normality of standard ferrous ammonium sulphate solution, and

V= Volume in ml of the sample taken for the test.

14.5 Precision and Accuracy - The method is quite precise and may be used in a wide variety of waters even though the back titration is less than 1ml. For most organic compounds, the oxidation is 95 to 100 per cent of the theoretical value. Using the silver sulphate catalyst, short straight chain alcohols and acids are oxidized to the extent of 85 to 95 per cent or better. Benzene, toluene and pyridine are not oxidized by either procedure.

15. Total Dissolved solids (Inorganic):-

15.1 Procedure - Determine total dissolved solids as prescribed in 7 of IS : 2488 (part III) 1968 (Methods of sampling and test for industrial effluents).

15.2 Calculation :-

$$\text{Total dissolved solids (inorganic), mg/1} = \frac{1000M}{V}$$

Where

M= mass in mg of the ignited residue, and

V =Volume in ml of the sample originally taken for the test,

16 Chloride:-

16.0 General - Chloride is one of the major anions in water and sewage. The salty taste produced by chloride concentrations is variable and dependent on the chemical composition of the water. Some waters containing 250mg/l chloride may evidence a detectable salty taste with sodium ions. On the other hand, the typical salty taste may be absent in waters containing as much chloride as 1000mg/l when there is a predominance of calcium and magnesium ions.

A high chloride content also exerts a deleterious effect on metallic pipes and structures, as well as on agricultural plants.

16.1 Selection of method - Two methods are presented for the determination of chloride in potable water. since the two methods are similar in most respects, selection is largely a matter of preference. Where colour offers a problem in the water sample, the determination can be performed potentiometrically.

16.A Argentometric Method :-

16.A.1 Principle - In a neutral or slightly alkaline solution, potassium chromate can indicate the end point of the silver nitrate titration of chloride. Silver chloride is quantitatively precipitated before red silver chromate is formed.

16.A.2 Interference - Substances in amounts normally found in potable waters will not interfere. Bromide, iodide and cyanide register as equivalent chloride concentrations, sulfide, thiosulfate and sulfite ions interfere. However, sulfite can be removed by treatment with hydrogen peroxide in a neutral solution, while sulfide and thiosulphate can be removed by treatment with hydrogen peroxide in alkaline solution. Orthophosphate in excess of 25 mg/l interferes by precipitation as silver phosphate. Iron in excess of 10mg/l will interfere by masking the end point.

16.A.3 Reagents:-

16A.3.1 Chloride-free water - If necessary, use redistilled or deionized distilled water.

16A.3.2 Potassium chromate indicator solution - Dissolve 50g K2Cr04 in a little distilled water. Add silver nitrate solution until a definite red precipitate is formed. Allow to stand 12 hr, filter, and dilute the filtrate to one litre with distilled water.

16A.3.3 Standard Silver nitrate titrant, 0.0141N - Dissolve 2.395g AgNO3 in distilled water and dilute to 1,000ml. Standardize against 0.0141 N Nacl by means of the procedure described below. Store in a brown bottle standard silver nitrate solution, exactly 0.0141 N, is equivalent to 500 µg Cl per 1.00ml.

16A.3.4 Standard Sodium Chloride 0.0141N - Dissolve 824.1mg NaCl (dried at 140ºC) in chloride-free water and dilute to 1000ml, 1.0ml = 500µg Cl.

16A.3.5 Special Reagents for removal of interference.

16A.3.5.1 Aluminium hydroxide suspension - Dissolve

125g aluminium potassium sulfate or aluminium ammonium sulfate, $AlK(SO_4)2.12H_2O$, or $AlNH_4(SO_4)2.12H_2O$, in one litre of distilled water. Warm to 60ºC and add 55ml Conc.NH4OH slowly with stirring. After allowing it to stand about 1hr, transfer the mixture to a large bottle and wash the precipitate by successive additions, with thorough mixing, and decantations of distilled water, until free from chloride. When freshly prepared, the suspension occupies a volume of approximately one litre.

16.A.3.5.2 Phenolphthalein indicator solution.

16.A.3.5.3 Sodium hydroxide, 1N.

16.A.3.5.4 Sulfuric acid, IN.

16.A.3.5.5 Hydrogen peroxide, 30%.

16.A.4 Procedure:-

16.A.4.1 Sample preparation - Use a 100ml sample or a suitable aliquat diluted to 100ml. If the sample is highly coloured, add 3ml Al(OH)3 suspension, mix, allow to settle, filter, wash and combine filtrate and washing. If sulfide, sulfite, or thiosulfate is present, make the water alkaline to phenolphthalein with sodium hydroxide solution. Add 1ml H_2O_2 and stir. Neutralize with sulphuric acid.

16A.4.2 Titration - Titrate samples in the pH range 7-10 directly. Adjust samples not in this range with sulphuric acid or sodium hydroxide solution. Add 1.0ml K_2CrO_4 indicator solution. Titrate with standard silver nitrate titrant to a pinkish yellow end point. The means of consistent end-point detection are left to the individual analyst.

Standardize the silver nitrate titrant and establish the reagent blank value by the titration method outlined above. A blank of 0.2 to 0.3ml is usual for the method.

16.A.5. Calculation :

$$mg/l\ Cl = \frac{(A-B) \times N \times 35{,}450}{ml\ Sample}$$

Where A = ml titration for Sample,

B= ml titration for blank, and

N= normality of Ag NO_3

mg/l NaCl = mg / l Cl x 1.65

16.A.6 Precision and Accuracy:-

A synthetic unknown sample containing 241mg/l chloride, 108mg/l Ca, 82mg/l Mg, 3.1mg/l K, 19.9mg/l Na, 1.1mg/l nitrate N, 250 µg/l nitrate N, 259 mg /l Sulfate, and 42.5mg/l total alkalinity (contributed by $NaHCO_3$) in distilled water was determined by the argentometric method with a relative standard deviation of 4.2% and a retative error of 1.7% in 41 laboratories.

16.B Mercuric Nitrate Method:-

16.B.1 Principle – Chloride can be titrated with mercuric nitrate because of the formation of soluble, slightly dissociated mercuric chloride. In the pH range 2.3-2.8, diphenyl carbazone indicates the end point of this titration by formation of a purple complex with the excess mercuric ions.

The error in titration is about 1% of the volume of titrant used per change of 0.1 pH unit in the pH range 2.1 - 2.8. Since exact pH adjustment is not feasible except by use of a pH meter, it is felt that keeping within a range of ± 0.1 pH unit is sufficient for most water analysis. Therefore, in this method, a specific mixture of nitric acid and diphenylcarbazone is added to a water sample, automatically adjusting the pH of most potable waters to pH 2.5 ± 0.1. A third substance in this alcoholic mixture, xylene cyanal FF, is used as a pH indicator and as a background colour to facilitate end-point detection. The introduction of 10mg sodium bicarbonate to both the blank and the standard titration provides a pH of 2.5±0.1 when 1.0ml indicator acidifier reagent is added. Increasing the strength of the titrant and modifying the indicator mixture enable determination of the higher chloride concentrations common in wastewater.

16.B.2 Interference - Bromide and iodide are titrated with mercuric nitrate in the same manner as chloride. Chromate, ferric, and sulfite ions interfere when present in excess of 10 mg/l.

16.B.3 Reagents:-

(a) Standard sodium chloride, 0.0141N

(b) Nitric acid, 0.I N.

(c) Sodium hydroxide 0.1N.

(d) Reagents for low-chloride titrations

(i) Indicator - Acidifier Reagent - The nitric acid concentration of this reagent is an important factor in the success of the determination and can be varied as indicated in (1) or (2) to suit the alkalinity range of the sample being titrated. Reagent (1) contains sufficient nitric acid to neutralize a total alkalinity of 150 mg /l as $CaCO_3$ to the proper pH in a 100ml sample.

Dissolve, in the order named, 250mg S-diphenyl carbazone, 4.0ml conc. nitric acid, and 30mg xylene cyanol FF in 100ml of 95% ethyl alcohol or isopropyl alcohol. Store in a dark bottle in a refrigerator. This reagent is not stable indefinitely. Deterioration causes a slow end point and high results. In as much as pH control plays a critical role in this method, adjust the pH of highly alkaline or acid samples to 2.5 $\pm$ 0.1 with 0.1N nitric acid or sodium hydroxide, not with sodium bicarbonate. Use a pH meter with a non-chloride type of reference electrode for the pH adjustment. If only the usual chloride-type reference electrode is available for the pH adjustment, determine the amount of acid or alkali required to achieve a pH of 2.5 $\pm$ 0.1 and discard this particular sample portion. Then treat a separate sample portion with the determined amount of acid or alkali and continue the analysis to its prescribed end. Under these circumstances, omit the nitric acid from the indicator reagent to maintain the proper sample pH. Alternatively vary the nitric acid concentration of the indicator-acidifier reagent to accommodate conditions wherein water samples of very high or very low alkalinity are being analysed.

(ii) Standard Mercuric Nitrate Titrant, 0.0141N - Dissolve 2.3g Hg $(NO_3)2\ H_2O$ in 100 ml distilled water containing 0.25ml conc. HNO_3. Dilute to just under one litre. Perform a preliminary standardization by following the procedure described in 16 A.4.1. Use replicates containing 5.0ml standard Nacl solution and 10mg $NaHCO_3$ diluted to 100ml with distilled water. Adjust the mercuric nitrate titrant to exactly 0.0141 N and perform a final standardization. Store away from the light in a dark bottle. Standard mercuric nitrate titrant, exactly 0.0141N, is equivalent to 500µg Cl per 1.00ml.

(e) **Reagents for high-chloride titrations :-**

(i) Mixed indicator Reagent - Dissolve 5g diphenylcarbazone powder and 0.5g bromphenol blue powder in 750ml 95% ethyl or isopropyl alcohol and dilute to one litre with ethyl or isopropyl alcohol.

(ii) Strong standard mercuric nitrate titrant. 0.141N.

Dissolve 25g Hg $(NO_3)2.H_2O$ in 900ml distilled water containing 5.0 ml conc. HNO_3. Dilute to just under one litre, and perform a preliminary standardization by following the procedure described in 16.A.4.2. Use replicates containing 25.00ml standard NaCl solution and 25ml distilled water. Adjust the titrant to exactly 0.141N and perform a final standardization. The chloride equivalence of the titrant is 5.00mg per 1.0ml.

16.B.4. Procedure:-

16.B.4.1 Titration of low-chloride concentrations prevailing in drinking water. Use a 100ml sample or smaller aliquot so that the chloride content is less than 10mg. Add 1.0ml of indicator-acidifier reagent to the sample. (The colour of the solution should be green-blue at this point). A light green indicates a pH of less than 2.0, a pure blue indicates a pH of more than 3.8. For most potable waters, the pH after this addition will be 2.5 ± 0.1. when highly alkaline or acid waters are encountered, a preliminary pH adjustment to about pH 8 will be necessary before the indicator-acidifier reagent is added. Titrate the treated sample with 0.0141 N mercuric nitrate titrant to a definite purple end point. The solution will turn from green-blue to a blue a few drops from the end point. Determine the blank by titration of 100 ml distilled water containing 10 mg $NaHCO_3$

16.B.4.2 Titration of high-Chloride concentrations - Place 50.0ml sample in a 150ml beaker (5.00ml sample may be used when more than 5ml titrant are needed). Add approximately 0.5ml mixed indicator reagent and mix well. The colour should be purple. Add 0.1N HNO_3 dropwise until the colour just turns yellow. Titrate with 0. 141N mercuric nitrate titrant to the first permanent dark purple. Titrate a distilled water blank using the same procedure.

16.B.5 Calculation.

$$mg / l\ Cl = \frac{(A-B) \times N \times 35{,}450}{ml\ Sample}$$

Where A = ml titration for sample

B =ml titration for blank, and N = normality of Hg $(NO_3)2$.

mg /l NaCl = mg /l Cl x 1.65

16.B.6 Precision and Accuracy - A synthetic unknown sample containing 241 mg/l Chloride, 108mg /l ca, 82mg/l Mg, 3.1 mg /lK, 13.9 mg Na, 1.1 mg / l nitrate N. 250 µg /l nitrite N, 259 mg/l sulfate, and 42.5mg/l total alkalinity (contributed by $NaHCO_3$) in distilled water was determined by the mercuric metric method, with a relative standard deviation of 3.3% and a relative error of 2.9% in 10 laboratories.

References

1. IDF (1979). Definitions of Pasteurization, Sterilization and UHT Treatment as applicable to milk and milk products. Recommendations of Commission D of the International Dairy Federation, Brussels.
2. IDF Bulletin (1986). Monograph on "Pasteurized Milk", International Dairy Federation Bulletin No. 200, 9.
3. Indian Standards Institution (1960). Methods of test Dairy Part I : Rapid Examination of Milk 1S : 1479, New Delhi.
4. Indian Standards Institution (1964). Specifications for Ice-cream 1S : 2802, New Delhi.
5. Indian Standards Institution (1964). Layout Plan for Dairy Laboratories 1S : 2981, New Delhi.
6. Indian Standards Institution (1964). Hard Cheese, Processed Cheese and Processed Cheese Spread, 1S : 2765, New Delhi.
7. Indian Standards Institution (1966). Methods of Sampling and Test for Butter 1S : 3507, New Delhi.
8. Indian Standards Institution (1966). Methods of Sampling and Test for Cream 1S : 3509, New Delhi.
9. Indian Standards Institutions (1967). Specification for sterilized Milk 1S : 4238, New Delhi.
10. IS : 4251 (1967), (Reaffirmed 1977). Indian Standards quality tolerances for water for processed food industry Bureau of Indian Standards, New Delhi.
11. IS : 4883 (1968). Indian Standards Specifications for Khoa, Indian Standards Institution, New Delhi.
12. Indian Standards Institution (1968). Specification for Infant Milk Foods IS : 1547, New Delhi.
13. Indian Standards Institutions (1969). Chhana, IS : 5162, New Delhi.
14. Indian Standards Institution (1969). Cleaning & Sanitization of Dairy Equipment IS : 5253, New Delhi.
15. IS : 5550 - (1970). Indian Standard Specifications of Burfi Bureau of Indian Standards, New Delhi.
16. Indian Standards Institution (1971). Code for Pasteurization of Milk IS : 6397, New Delhi.

17. Indian Standards Institution (1973). Specifications for Condensed Milk IS : 1166, New Delhi.

18. IS : 7035(1973). Indian Standards Specifications for fermented Milk Products, Indian Standards Institution, New Delhi.

19. Indian Standards Institution (1975). Specifications for Milk Powder, 1S : 1165, New Delhi.

20. Indian Standards Institution (1977). Part III Bacteriological Analysis of Milk 1S : 1479, New Delhi.

21. 1S1 (1981). 1S1 Handbook of Food Analysis SP : Part XI, Dairy Products Indian Standards Institution Manak Bhawan, 9, Bahadur Shah Zafar, Marg New Delhi, 110002.

22. Iya, K.K. (1962). In : Milk Hygiene, WHO Monograph Series No. 48 FAO / WHO, Geneva, P. 620.

Index

V

W

Y

Z